Commentary

DoD Needs to Better Incorporate Adaptation into Planning and Collaboration at Overseas Installations – GAO-18-206, November 2017

DOD installations overseas have experienced operational and budgetary risks posed by weather effects associated with climate change impacts at the military services' installations in each of DOD's geographic combatant commands. This book demonstrates the very real impact of global warming and the need to strengthen the resiliency of infrastructure used by DOD. This book identifies recommendations for executive action.

Climate change deniers can stick their heads in the sand and pretend climate change is not real, but this book explains the science behind what is happening and clearly documents what should otherwise be obvious. In addition to sea level will rise, the DoD installations have been damaged by extreme weather events. To make matters worse, the land is sinking in some places and groundwater is being impacted. Global weather conditions are getting worse so even if the sea level rises slower than expected, hurricane storm surge and increased rainfall will wreak havoc with the electric grid, natural gas pipelines, municipal water and sewage treatment plants. Climate change will also result in humanitarian stressors for the DoD to deal with after a natural disaster. That said, we can expect wildfires in parts of the country not known for wildfires and flooding everywhere along the coasts. In our books "Sea Level Rise Maps" we show where coastal flooding will begin starting as early as the year 2030. We can expect to lose more than fancy beach houses. At least 30 major airports in cities like Boston, New York's JFK and LaGuardia, Newark, Philadelphia, Miami, San Diego and San Francisco will all be completely underwater by the year 2100 – and there is nothing anyone can do about it. Even if we reduce our carbon footprint to zero today, the process has already begun.

Why buy a book you can download for free? We print this book so you don't have to.

First you gotta find a good clean (legible) copy and make sure it's the latest version (not always easy). If you find a good copy, you could print it using a network printer you share with 100 other people (typically its either out of paper or toner). If it's just a 10-page document, no problem, but if it's 250-pages, you will need to punch 3 holes in all those pages and put it in a 3-ring binder. Takes at least an hour.

It's much more cost-effective to just order the latest version from Amazon.com
https://usgovpub.com

Other books we print that are available on Amazon.com:

Fourth National Climate Assessment – Volume I

Fourth National Climate Assessment – Volume II

The Impact of Sea-Level Rise and Climate Change on Department of Defense Installations on Atolls in the Pacific Ocean (RC-2334) – 2017

Sea Level Rise Maps – U.S. East Coast – 2100

Sea Level Rise Maps – U.S. Gulf Coast – 2040-2100

Home Builder's Guide to Construction in Wildfire Zones

Home Builder's Guide to Coastal Construction

Protecting Building Utility Systems from Flood Damage

Local Officials Guide for Coastal Construction

FEMA Incident Management Handbook

FEMA Incident Action Planning Guide

Federal Acquisition Regulation (FAR)

GAO Financial Audit Manual (GAO FAM)

GAO Government Auditing Standards (Yellow Book) GAO-18-568G

GAO Standards for Internal Control in the Federal Government (Green Book) GAO-14-704G

GAO Internal Control Management and Evaluation Tool GAO-01-1008G

OMB A-123   Management's Responsibility for Enterprise Risk Management and Internal Control

Contract Attorneys Deskbook

Fiscal Law Deskbook

DOD Energy Manager's Handbook

United States Government Accountability Office
Report to Congressional Requesters

November 2017

CLIMATE CHANGE ADAPTATION

DOD Needs to Better Incorporate Adaptation into Planning and Collaboration at Overseas Installations

GAO Highlights

Highlights of GAO-18-206, a report to congressional requesters

November 2017

CLIMATE CHANGE ADAPTATION

DOD Needs to Better Incorporate Adaptation into Planning and Collaboration at Overseas Installations

Why GAO Did This Study

According to DOD, climate change will have serious implications on the ability to maintain infrastructure and ensure military readiness. DOD has identified risks posed by climate change and begun to integrate adaptation in guidance. GAO was asked to assess DOD's actions to adapt overseas infrastructure to the expected challenges of climate change.

GAO examined the extent to which DOD (1) identified operational and budgetary risks posed by weather effects associated with climate change on overseas infrastructure; (2) collected data to effectively manage risks to infrastructure; (3) integrated climate change adaptation into planning and design efforts; and (4) collaborated with host nations on adapting infrastructure and sharing costs. GAO reviewed DOD data and documents on climate change, planning, and cost-sharing and visited or contacted a nongeneralizable sample of 45 overseas installations reporting climate change impacts.

What GAO Recommends

GAO is making six recommendations, including that DOD require overseas installations to systematically track costs associated with climate impacts; re-administer its vulnerability assessment survey to include all relevant sites; integrate climate change adaptation into relevant standards; and include climate change adaptation in host-nation agreements. DOD nonconcurred with two recommendations and partially concurred with four. GAO recognizes DOD's efforts to review its climate-related policies, but continues to believe its recommendations are valid, as discussed in this report.

View GAO-18-206. For more information, contact Brian J. Lepore at (202) 512-4523 or leporeb@gao.gov.

What GAO Found

The expected impacts of weather effects associated with climate change pose operational and budgetary risks to overseas infrastructure according to the Department of Defense (DOD), but DOD does not consistently track the impacts' estimated costs. Operational risks (including interruptions to training, testing, and missions) and budgetary risks (including costs of repairing damages) are linked to these impacts. However, installations inconsistently track these costs because there is no requirement for such tracking. Without a requirement to systematically track such costs, DOD will not have the information it needs to integrate climate-related impact resource considerations into future budgets.

Severe Erosion at a Department of Defense (DOD) Munitions Storage Complex in the Pacific

According to installation officials, this munitions complex is threatened by erosion and flooding caused by increasingly frequent and intense rain events. In previous work examining climate change impacts on DOD infrastructure, GAO found that it is not possible to link any individual weather event to climate change. However, these events provide insight into the potential climate-related vulnerabilities faced by DOD.

Source: DOD. | GAO-18-206

DOD surveyed overseas installations on their vulnerability to the operational and budgetary risks of weather effects associated with climate change, but the approach used to gather survey data on the impacts that cause these risks was incomplete and not comprehensive. Specifically, DOD exempted dozens of overseas sites from completing the vulnerability assessment, and did not include key national security sites. As a result, DOD did not obtain information on risks posed by weather effects associated with climate change at many key overseas installations, which is critical for managing such risks at these locations.

While the military services have begun to integrate climate change adaptation into installations' plans and project designs, this integration has been limited. For example, only about one-third of the plans that GAO reviewed addressed climate change adaptation. Similarly, projects GAO discussed with DOD officials were rarely designed to include climate change adaptation. This is due to the inconsistent inclusion of climate change adaptation in training and design standards for installation planners and engineers. As a result, planners and engineers do not have the information needed to ensure that climate change-related risks are addressed in installation plans and project designs.

DOD collaborates with host nations at both the national and installation level, but cost sharing agreements and other collaboration efforts generally do not include climate change adaptation. Without more fully including adaptation into its agreements with host nations, DOD may miss opportunities to increase the resilience of host-nation-built infrastructure at overseas installations to risks posed by the weather effects associated with climate change.

United States Government Accountability Office

Contents

Letter		1
	Background	8
	DOD Installations Have Identified Operational and Budgetary Risks Posed by the Impacts of Weather Effects Associated with Climate Change, but Do Not Consistently Track the Costs of These Impacts	18
	DOD Has Collected Incomplete Data on Overseas Climate Impacts, Which Creates Challenges for Effectively Managing Climate Risks	29
	DOD Has Included Climate Change Adaptation in Infrastructure Guidance, but Integration into Installation-Level Planning Is Limited and Design Standards Have Not Been Updated	35
	Climate Change Adaptation Is Generally Not Included in DOD's Collaboration with Host Nations at the National or Installation Level	45
	Conclusions	52
	Recommendations for Executive Action	54
	Agency Comments and Our Evaluation	55
Appendix I	Scope and Methodology	63
Appendix II	Comments from the Department of Defense	72
Appendix III	GAO Contact and Staff Acknowledgments	75
Related GAO Products		76

Tables

	Table 1: Seven Potential and Observed Climate Change Impacts on Department of Defense (DOD) Infrastructure and Operations	10
	Table 2: Number of Department of Defense (DOD) Overseas Locations Exempted from the Survey	32
	Table 3: Department of Defense (DOD) Installations We Visited or Contacted	69

Figures

Figure 1: Flooding on Roads and in Facilities (Left) and Mud Covering the Ramp of a Runway (Right) at a Department of Defense Installation in Europe — 20

Figure 2: Water Level of a Canal Adjacent to Ammunition Loading Docks at a Department of Defense Installation in Europe — 21

Figure 3: Security Checkpoint in Non-flood Conditions (Left) and During Flooding in 2014 (Right) at a Department of Defense Installation in the Pacific — 22

Figure 4: Flooding (Left) and Erosion (Right) at a Department of Defense Installation in Africa — 23

Figure 5: Helicopter Damaged by an Extreme Wind Storm at a Department of Defense Installation in Europe — 24

Figure 6: Erosion along Roads at a Department of Defense Training Range in the Pacific — 26

Figure 7: Hillside Impacted by Erosion at a Department of Defense Installation in Europe — 37

Figure 8: Movement of Sea Turtles across a Department of Defense (DOD) Beach, Indicated by Arrows (Left) and Sea Turtle Hatchlings (Right), at a DOD Installation in the Pacific — 38

Figure 9: 2010 and 2014 Rainstorms Overwhelmed the Drainage System at a Department of Defense Installation in the Pacific — 43

Figure 10: Damaged Seawall at a Department of Defense Ammunition Depot in the Pacific — 47

Abbreviations

Roadmap	2014 Climate Change Adaptation Roadmap
2016 Sustainability Plan	2016 Strategic Sustainability Performance Plan
DOD	Department of Defense
State	Department of State
Survey	Screening Level Vulnerability Assessment Survey

This is a work of the U.S. government and is not subject to copyright protection in the United States. The published product may be reproduced and distributed in its entirety without further permission from GAO. However, because this work may contain copyrighted images or other material, permission from the copyright holder may be necessary if you wish to reproduce this material separately.

GAO

U.S. GOVERNMENT ACCOUNTABILITY OFFICE

441 G St. N.W.
Washington, DC 20548

November 13, 2017

Congressional Requesters

The Department of Defense (DOD) manages a global real-estate portfolio that includes almost 600 overseas sites with a plant replacement value DOD estimates at about $158 billion dollars. DOD uses an extensive portfolio of overseas infrastructure—including facilities owned by host nations—that extends across each of its geographic combatant commands and is critical to maintaining military readiness.[1] DOD guidance states that the U.S. foreign and overseas posture is the fundamental enabler of U.S. defense activities and military operations overseas and is also central to defining and communicating U.S. strategic interests to allies, partners, and adversaries.[2] Since 2010, DOD has identified climate change as a threat to its operations and installations (i.e., operational risks) and has reported that it needs to adapt its infrastructure to the risks posed by climate change.[3] Further, as climate

---

[1]For the purposes of this report, we define infrastructure as all buildings and permanent installations necessary for the support, redeployment, and operations of forces (e.g., barracks, headquarters, airfields, communications facilities, stores, port installations, and maintenance stations). Infrastructure includes utility systems; training and testing ranges and areas; and transportation systems (e.g., roads and bridges). It also includes any built or natural infrastructure outside of a facility (e.g., utility lines or barrier islands, respectively) that DOD officials considered in adaptation planning. Also, DOD has six geographic combatant commands, each with defined areas of operation and a distinct regional military focus. They are the U.S. Africa Command, U.S. Central Command, U.S. European Command, U.S. Northern Command, U.S. Pacific Command, and U.S. Southern Command. The geographic combatant commands provide unity of command over all the U.S. forces in a specific region.

[2]DOD Instruction 3000.12, *Management of U.S. Global Defense Posture (GDP)* (May 6, 2016).

[3]The *DOD Dictionary of Military and Associated Terms* (August 2017) defines climate change as variations in average weather conditions that persist over multiple decades or longer that encompass increases and decreases in temperature, shifts in precipitation, and changing risk of certain types of severe weather events. In our previous work on climate change, we defined the term adaptation as adjustments to natural or human systems in response to actual or expected climate change; adaptation is synonymous with enhancing resilience. See GAO, *High-Risk Series: Progress on Many High-Risk Areas, While Substantial Efforts Needed on Others*, GAO-17-317 (Washington, D.C.: Feb. 15, 2017). Adaptation includes considerations of climate change, such as whether or not specific adaptation actions are necessary, based on risk. Climate change adaptation differs from mitigation, which is focused on reducing emissions. Operational risks include interruptions or delays in base or mission operations (e.g., base personnel not able to work on an installation due to a climate impact, such as a severe storm, or rescheduling training range activities due to flooding).

change impacts damage infrastructure, requiring repairs, these impacts result in costs (i.e., budgetary risks) to the department.[4] Scientific projections and observations indicate that climate change includes increasing temperatures, rising sea levels, and other gradual changes, as well as the potential for increases in the frequency and severity of extreme weather events.[5]

Climate change is considered by many to be a complex, crosscutting issue that poses risks to many environmental and economic systems.[6] In February 2013, we placed the federal government's fiscal exposure to climate change on our High Risk List.[7] As part of our work in this high-risk area, we found in May 2014 that DOD had not fully incorporated climate change adaptation into its domestic infrastructure planning and investment efforts, planned for the use of vulnerability assessment data, or provided installation planners with needed information on how to adapt to climate change impacts.[8] We made three recommendations to address these findings. As of June 2017, DOD had made progress in a number of areas that are important to adapting its domestic infrastructure to these

---

[4]Budgetary risks include the use of funding to prepare for, or recover from, climate impacts (e.g., the cost of overtime required to set up sandbags in anticipation of flooding or repair roofs destroyed during a severe wind storm).

[5]According to the National Academies of Sciences, Engineering, and Medicine, extreme weather has affected human society since the beginning of recorded history. Also, the National Academies report that humans' use of fossil fuel since the start of the Industrial Revolution has begun to modify the Earth's climate. Further, according to the United Nations Intergovernmental Panel on Climate Change, changes in climate have caused impacts on natural and human systems on all continents and across the oceans. The United Nations panel reports that impacts from recent climate-related extremes—such as heat waves, droughts, floods, cyclones, and wildfires—reveal significant vulnerability and exposure of some ecosystems and many human systems to current climate variability. This report states that impacts of such climate-related extremes include, among others, alteration of ecosystems and damage to infrastructure. The National Academies Press, *Attribution of Extreme Weather Events in the Context of Climate Change* (Washington, D.C.: 2016) and Intergovernmental Panel on Climate Change, 2014: Summary for policymakers. In: *Climate Change 2014: Impacts, Adaptation, and Vulnerability. Part A: Global and Sectoral Aspects. Contribution of Working Group II to the Fifth Assessment Report of the Intergovernmental Panel on Climate Change.*

[6]GAO, *High-Risk Series: An Update*, GAO-15-290 (Washington, D.C.: Feb. 11, 2015).

[7]GAO, *High-Risk Series: An Update*, GAO-13-283 (Washington, D.C.: Feb. 14, 2013). See also GAO-17-317.

[8]GAO, *Climate Change Adaptation: DOD Can Improve Infrastructure Planning and Processes to Better Account for Potential Impacts*, GAO-14-446 (Washington, D.C.: May 30, 2014).

impacts, had implemented one recommendation, and had taken steps toward implementing the remaining two recommendations. For example, in regard to our recommendation to facilitate the efforts of installation planners to efficiently implement the requirements of DOD guidance on climate change adaptation for infrastructure, DOD defined key terms in subsequent guidance and provided information about projected sea level change and associated impacts for individual installations. However, the department has not yet provided these planners with projections for the full set of expected impacts of weather effects associated with climate change. Also, in regard to our recommendation to clarify instructions for the comparison of potential military construction projects to include consideration of climate change adaptation, the Army has considered climate change adaptation as a project component that may be needed to address potential climate change impacts on infrastructure for at least two domestic projects. However, DOD has not provided us with evidence that the department's components have clarified instructions associated with the processes used to compare potential military construction projects for approval and funding.

We were asked to assess how DOD is adapting its overseas infrastructure to climate change. This report assesses the extent to which DOD has (1) identified the operational and budgetary risks posed by weather effects associated with climate change on infrastructure used by DOD overseas; (2) collected data to effectively manage the operational and budgetary risks of weather effects associated with climate change to this infrastructure; (3) integrated adaptation to weather effects associated with climate change into its installation planning and project design efforts; and (4) collaborated with host nations on adapting infrastructure used by DOD to increase resiliency to the impacts of weather effects associated with climate change and shared costs for any needed adaptation. Overseas, DOD executes missions using infrastructure the department owns and leases, as well as infrastructure owned by host nations. For that reason, in this report, we focus on all infrastructure used by DOD overseas.

To examine the extent to which DOD has identified the operational and budgetary risks posed by weather effects associated with climate change on infrastructure used by DOD overseas, we reviewed data collected by DOD's Screening Level Vulnerability Assessment Survey (survey) from 2013 to 2015. Using these data, we developed a nongeneralizable sample of 45 overseas installations. We interviewed military service officials and collected documentation on the observed impacts of extreme weather events and climate change, as well as associated costs, at these installations. To develop the sample, we identified installations that reported at least one of seven effects that, according to a DOD survey, are associated with climate change.[9] To select locations to visit in person, we assessed each installation in our sample against a number of factors, including the number and type of climate change impacts reported for the installation; the military service located at the installation; and the country in which the installation and its associated sites were located. These installations were spread across 22 countries in each of the six geographic combatant commands' areas of responsibility. We visited 14 of the 45 installations within the U.S. European Command and U.S. Pacific Command to observe both the physical impacts of extreme weather events and climate change on infrastructure and adaptation or resilience measures taken or planned at the installation level. During our site visits, we interviewed installation officials about observed impacts to infrastructure and collected key documentation describing impacts. Although the information we collected is not representative of all DOD installations overseas, the risks and impacts these installations identified—and the adaptation efforts they have taken—provide valuable insights. To gather additional information on the impacts observed by DOD personnel, and associated costs, we interviewed and reviewed documentation from DOD officials, including those in the Office of the Under Secretary of Defense, the geographic combatant commands, and the military services' headquarters and installations in our sample. We also interviewed installation officials about the types of costs associated with the risks posed by weather effects associated with climate change, such as facility maintenance and repair costs, and how often DOD installations track these costs. We compared what installation officials told us about tracking these costs with DOD Directive 4715.21, which requires DOD components to incorporate climate change resource considerations

---

[9]DOD's survey identified seven effects commonly associated with climate change: flooding due to storm surge, flooding due to non-storm surge events (rain, snow, sleet, ice, and river overflow), extreme temperatures (both hot and cold), wind, drought, wildfire, and changes in mean sea level.

related to adapting built and natural infrastructure to potential or observed climate change impacts into installation-level planning efforts.[10]

To examine the extent to which DOD has collected data to effectively manage the operational or budgetary risks of weather effects associated with climate change to overseas infrastructure, we reviewed DOD guidance, to include DOD Directive 4715.21, to understand the military services' roles and responsibilities for assessing climate change impacts on infrastructure. To determine DOD's goals for conducting vulnerability assessments, we also reviewed DOD's 2010 and 2014 *Quadrennial Defense Reviews* and the 2014 *Climate Change Adaptation Roadmap (Roadmap)*.[11] In addition, we reviewed DOD's 2016 *Strategic Sustainability Performance Plan (2016 Sustainability Plan)*, which addresses the department's approach to the management of the risks posed by climate change.[12] We also reviewed guidance from the Office of the Secretary of Defense that accompanied the administration of DOD's survey, which required that the military services survey their installations about the risks posed by weather effects associated with climate change, and survey instructions that the department provided to the military services, along with best practices for conducting surveys.[13] Finally, we reviewed DOD's 2016 Enduring Locations Master List to identify overseas infrastructure of particular significance to DOD missions.

---

[10] DOD Directive 4715.21, *Climate Change Adaptation and Resilience* (Jan. 14, 2016). DOD Directive 4715.21 was issued in accordance with the direction in Executive Order 13653. On March 28, 2017, the Presidential Executive Order on Promoting Energy Independence and Economic Growth (Executive Order 13783) rescinded Executive Order 13653. This rescinded executive order stipulated that, among other things, each agency was to develop or continue to develop, implement, and update comprehensive adaptation plans that integrate consideration of climate change into agency operations and overall mission objectives. According to an official from the Office of the Secretary of Defense, as of September 2017, DOD was working to determine the course of action the department will take with regard to its directive to comply with Executive Order 13783.

[11] DOD, *Quadrennial Defense Review Report* (Feb. 1, 2010); DOD, *Quadrennial Defense Review 2014* (Mar. 4, 2014); and, DOD, *2014 Climate Change Adaptation Roadmap* (Alexandria, VA: June 2014) (hereinafter cited as DOD, 2014 *Roadmap*).

[12] DOD, *Strategic Sustainability Performance Plan FY 2016* (Sept. 7, 2016).

[13] Through the survey, DOD installations could report the following climate change impacts: drought, extreme temperatures (hot or cold), flooding and other impacts due to non-storm surge events, flooding due to storm surge, implications of increased mean sea level, wildfire, and wind. For a discussion of best practices for conducting surveys, see GAO, *The Democratic Republic of the Congo: Information on the Rate of Sexual Violence in War-Torn Eastern DRC and Adjoining Countries*, GAO-11-702 (Washington, D.C.: July 13, 2011).

To examine the extent to which DOD has integrated adaptation to weather effects associated with climate change into its installation planning and project design efforts, we reviewed guidance requiring DOD components to integrate climate change adaptation into certain installation and infrastructure planning efforts contained in DOD's *Unified Facilities Criteria 2-100-01, Installation Master Planning* and DOD Instruction 4715.03, *Natural Resources Conservation Program*.[14] We also reviewed plans (i.e., installation master plans, natural resources management plans, and encroachment management plans) from the installations in our sample and assessed these plans against DOD policy on incorporating climate change adaption into installation planning efforts. In addition, we interviewed DOD officials at the military services' headquarters and at selected installations to determine the extent to which the services have implemented climate change adaptation efforts at the installation level. We also interviewed DOD officials about installation-level planning efforts and planned or completed climate change adaptation projects.

To examine the extent to which DOD has collaborated with host nations on adapting infrastructure used by DOD to increase installation resiliency to the impacts of weather effects associated with climate change and shared costs for any needed adaptation, we collected information from DOD and Department of State (State) officials on collaboration between DOD and host nations on climate change adaptation and cost-sharing activities. We interviewed DOD officials from the regional service components, the sub-unified commands, and the installations in our sample to learn more about collaboration and cost-sharing related to climate change adaptation at the installation level. We reviewed bilateral

---

[14]DOD's Unified Facilities Criteria 2-100-01, *Installation Master Planning* (May 15, 2012) states that installation planners can prepare a master plan that sustainably accommodates future change by incorporating current needs and mission requirements into a vision with clear goals and measurable objectives. The guidance further states that the military services' master planners are to understand, monitor, and adapt to, among other things, changing climatic conditions. We also reviewed DOD Instruction 4715.03, governing the department's Natural Resources Conservation Program on domestic installations. According to DOD officials, the military departments have chosen to use the instruction as guidance for their overseas installations' development of these plans and installations in our sample have used the instruction in this way. The instruction states that all DOD natural resources conservation programs shall be integrated with installation planning and programming. The guidance further states that for natural resources plans all DOD components are to utilize existing tools to assess the potential impacts of climate change to natural resources on DOD installations to the extent practicable and using the best science available. DOD Instruction 4715.03, *Natural Resources Conservation Program* (Mar. 18, 2011).

agreements between DOD and host-nation governments for selected installations in our sample to determine the extent to which these agreements include information on climate change adaptation, collaboration on climate change challenges, or cost-sharing related to climate change. We also reviewed State's 2016 Treaties in Force for international agreements between the United States and host nations in our sample, which include information on climate change adaptation. We compared this information with DOD's 2014 *Roadmap* and DOD Directive 4715.21 to learn about DOD's goals and requirements for collaboration with external stakeholders to address climate change challenges, and reviewed our prior work related to leading practices for collaboration.[15]

By discussing potential sites for review with military service officials and by reviewing relevant DOD reports and database characteristics, we determined that DOD's vulnerability assessment survey database was sufficiently reliable to use as part of our site selection methodology and to generate questions for data-gathering from sites visited or contacted. Also, by discussing the process by which the Office of the Secretary of Defense, military services, and Joint Staff selected survey sites, we determined that DOD's vulnerability assessment survey database was sufficiently reliable to assess the extent to which DOD effectively used the data to manage the operational and budgetary risks posed by weather effects associated with climate change. In addition, by reviewing relevant sites' data for any seeming outliers, we determined that DOD's Regionalized Sea Level Change Scenarios Database was sufficiently reliable to use as a source of data on which to base questions for sites visited or contacted and to illustrate cases in which installations may not be using available data in their installation planning or project design efforts. Further details on our scope and methodology can be found in appendix I.

We conducted this performance audit from May 2016 to November 2017 in accordance with generally accepted government auditing standards. Those standards require that we plan and perform the audit to obtain sufficient, appropriate evidence to provide a reasonable basis for our findings and conclusions based on our audit objectives. We believe that the evidence obtained provides a reasonable basis for our findings and conclusions based on our audit objectives.

---

[15]GAO, *Managing for Results: Key Considerations for Implementing Interagency Collaborative Mechanisms*, GAO-12-1022 (Washington, D.C.: Sept. 27, 2012).

Background

According to the National Research Council, although the exact details cannot be predicted with certainty, climate change poses serious risks to many of the physical and ecological systems upon which society depends.[16] Moreover, according to key scientific assessments, the impacts and costs of extreme events—such as floods, drought, and other events—will increase in significance as what are considered rare events become more common and intense because of climate change.[17] According to the National Academies of Sciences, Engineering, and Medicine, extreme events are directly traceable to loss of life, rising food and energy prices, increasing costs of disaster relief and insurance, fluctuations in property values, and concerns about national security.[18]

As such, a variety of climate change effects are expected to impact overseas regions where DOD owns infrastructure or uses host nations' infrastructure. Examples include a marked increase in high temperature extremes in Europe or an increase in heavy rain events that could impact Asia. In a draft report on the results of its Screening Level Vulnerability Assessment Survey (Survey Summary Report), the department identified

---

[16]The National Research Council is the principal operating agency of the National Academies of Sciences, Engineering, and Medicine for furnishing scientific and technical advice to governmental and other organizations. *See,* National Research Council, Committee on America's Climate Choices, *America's Climate Choices* (Washington, D.C.: 2011); National Research Council, *Climate Change: Evidence, Impacts, and Choices. Answers to common questions about the science of climate change* (Washington, D.C.: 2012).

[17]Jerry M. Melillo, Terese (T.C.) Richmond, and Gary W. Yohe, eds., Climate Change Impacts in the United States: The Third National Climate Assessment, (Washington, D.C.: U.S. Global Change Research Program, May 2014). and Intergovernmental Panel on Climate Change, 2014: Climate Change 2014: Impacts, Adaptation, and Vulnerability. Part A: Global and Sectoral Aspects. Contribution of Working Group II to the Fifth Assessment Report of the Intergovernmental Panel on Climate Change [Field, C.B., V.R. Barros, D.J. Dokken, K.J. Mach, M.D. Mastrandrea, T.E. Bilir, M. Chatterjee, K.L. Ebi, Y.O. Estrada, R.C. Genova, B. Girma, E.S. Kissel, A.N. Levy, S. MacCracken, P.R. Mastrandrea, and L.L. White (eds.)]. Cambridge University Press, Cambridge, United Kingdom and New York, NY, USA, 1132 pp.

[18]National Academies of Sciences, Engineering, and Medicine. 2016. Attribution of Extreme Weather Events in the Context of Climate Change. Washington, DC: The National Academies Press. doi: 10.17226/21852. According to the National Academies of Sciences, Engineering, and Medicine, the ability to attribute the causes of some extreme event types has advanced rapidly since the emergence of event attribution science a little more than a decade ago, while attribution of other event types remains challenging. Confidence in attribution of specific extreme events is highest for extreme heat and cold events, followed by hydrological drought and heavy precipitation. For example, for extreme heat and cold events in particular, changes in long-term mean conditions provide a basis for expecting that there also should be related changes in extreme conditions.

seven effects commonly associated with climate change, including: flooding due to storm surge, flooding due to non-storm surge events, extreme temperatures, wind, drought, wildfire, and changes in mean sea level.[19] DOD documentation states that observed effects from past severe weather events, such as flooding or wildfire, may be indicative of more frequent or more severe future conditions. Table 1 summarizes potential and observed examples of the seven effects commonly associated with climate change that DOD has documented.

---

[19]Office of the Secretary of Defense, draft Screening Level Vulnerability Assessment Survey Report (December 2016).

Table 1: Seven Potential and Observed Climate Change Impacts on Department of Defense (DOD) Infrastructure and Operations

Category	Potential climate change effect on weather events	Potential and observed impacts on DOD infrastructure and operations
Flooding due to storm surge events	Increased severity and frequency of storm surge flooding events	Coastal erosion (e.g., shoreline facilities), damage of coastal infrastructure (e.g., piers and utilities)
Flooding due to non-storm surge events	Increased severity and frequency of non-storm surge flooding events	Inland site inundation, infrastructure damage (e.g., training area facilities), training encroachment (e.g., excessive damage to maneuver training lands), storm water and wastewater disposal issues, shifting river flows
Extreme temperatures	Hot: Increased frequency of extreme hot days, permafrost thaw, seasonal weather shifts	Strained electricity supply, changing building cooling demand (e.g., impacting installation energy intensity and operating costs), training encroachment (e.g., more red and black flag days),[a] erosion and facility damage from thawing permafrost, water supply shortages, increased maintenance requirements for runways or roads
	Cold: Increased frequency of extreme cold days, seasonal weather shifts	Strained electricity supply, changing building heating demand (e.g., impacting installation energy intensity and operating costs), training encroachment, increased maintenance requirements for runways or roads
Wind	Stronger and more frequent wind events	Damage to above-ground electric/power infrastructure (e.g., power lines), roofs of buildings, and housing
Drought	Increased frequency of drought	Water supply shortages
Wildfire	Increased frequency of wildfire	Training encroachment (e.g., restrictions on types of ammunition used, halting or delaying training activities)
Changes in mean sea level	Increased severity and frequency of coastal flooding events	Coastal site damage due to erosion and inundation, water supply interruptions, wastewater disposal issues

Source: GAO analysis of the 2010 *Quadrennial Defense Review*, 2012 DOD *Climate Change Adaptation Roadmap (Roadmap)*, 2014 *Roadmap*, Fiscal Year 2015 *DOD Strategic Sustainability Performance Plan (Sustainability Plan)*, Fiscal Year 2016 *Sustainability Plan*, and the December 2016 draft of the *Screening Level Vulnerability Assessment Survey Report*. | GAO-18-206

[a]According to the U.S. Navy, red flag days are when strenuous exercise must be curtailed in hot weather for all personnel with fewer than 12 weeks of training; black flag days are when non-mission essential physical training and strenuous exercise must be suspended for all personnel.

In previous work examining climate change impacts on DOD infrastructure, we found that while it is not possible to link any individual weather event to climate change, these events provide insight into the potential climate-related vulnerabilities faced by DOD. We also found that, according to DOD installation-level officials, the department's facilities and infrastructure are vulnerable to climate change phenomena. Further, these officials recognized that climate change may make these types of phenomena more frequent or severe.[20]

---

[20]For a summary of our previous work on U.S. government climate change adaptation efforts and related recommendations for improvement of these efforts, see GAO-17-317. We have also included a list of related GAO products at the end of this report.

DOD Strategy and Policy Addressing Climate Change Impacts and Adaptation

Since 2010, DOD has—in key strategy documents and policy—cited the impacts that climate change is having, and is expected to have, on its infrastructure and operations, stressing the importance of adapting to these impacts in order to accomplish the department's mission. Selected strategy documents and policy documents, by year, include the following:

- **2010**: In its 2010 *Quadrennial Defense Review*, DOD stated that the department's operational readiness hinges on continued access to land, air, and sea training and test space. Further, DOD stated that it was developing policies and plans to manage the effects of climate change on its operating environment, missions, and facilities.[21]

- **2011**: In the 2011 *National Military Strategy*, DOD's characterization of the strategic environment included climate change as a potentially serious impact.[22]

- **2012**: In its fiscal year 2012 *Strategic Sustainability Performance Plan*, DOD stated that climate change can directly impact military installations and operations by limiting the availability and quality of training ranges and other lands needed for operations and by increasing impacts on infrastructure, such as flood and fire hazards and vulnerability of utilities.[23]

- **2013**: In its 2013 *Arctic Strategy*, DOD recognized that decreasing seasonal ice will increase access and activity in the region, potentially altering the security environment in which the department operates.[24]

- **2014**:
    - In its 2014 *Roadmap*, DOD stated that climate change will affect the department's ability to defend the nation and poses immediate risks to U.S. national security. The Roadmap focused on four lines of effort, including built and natural infrastructure. DOD stated that both built and natural infrastructure is necessary for successful

---

[21]DOD, *Quadrennial Defense Review Report* (Washington, D.C.: Feb. 1, 2010).

[22]Chairman of the Joint Chiefs of Staff, *The National Military Strategy of the United States of America 2011: Redefining America's Military Leadership* (Washington, D.C.: Feb. 11, 2011).

[23]DOD, *Strategic Sustainability Performance Plan FY 2012* (Sept. 20, 2012). In subsequent *Strategic Sustainability Performance Plans* from 2013 through 2016, DOD has continued to emphasize the serious impact that climate change will have on operations and infrastructure.

[24]DOD, *Arctic Strategy* (Washington, D.C.: November 2013).

mission preparedness and readiness. Specifically, built infrastructure serves as the staging platform for the department's national defense and humanitarian missions, and natural infrastructure supports military combat readiness by providing realistic combat conditions and vital resources to personnel.

- In its 2014 *Quadrennial Defense Review*, DOD stated that the impacts of climate change may undermine the capacity of installations to support training activities; that the department will complete a comprehensive assessment of all installations to assess the potential impacts of climate change on its missions and operational resiliency; and that the department will develop and implement plans to adapt, as required.[25]

- In its 2014 *Arctic Roadmap*, the Navy discussed the role that climate change plays in several national security arenas, such as energy security, the U.S. economy, and national sovereignty.[26]

- **2015:** The United States' *National Security Strategy* stated that climate change is one of the top strategic risks to our country's national interests, noting that climate change is an urgent and growing threat to national security. For example, increased sea levels and storm surges threaten coastal regions and infrastructure.[27]

- **2016:**

  - The 2016 *Sustainability Plan* states that climate change is a clear national security concern—affecting DOD today and forecasted to affect the department more in the future—and that climate change impacts can directly interfere with an installation's ability to carry out its mission. Further, the 2016 *Sustainability Plan* states that by incorporating aggressive consideration of the current and potential impacts of a changing climate in mission planning across the defense enterprise, DOD will become more sustainable.[28]

  - In its 2016 directive, *Climate Change Adaptation and Resilience*, DOD established policy that the department must be able to adapt current and future operations to address the impacts of climate

---

[25]DOD, *Quadrennial Defense Review 2014* (Washington, D.C.: Mar. 4, 2014).

[26]Chief of Naval Operations, *The United States Navy Arctic Roadmap for 2014 to 2030* (Washington, D.C.: February 2014).

[27]The White House, *National Security Strategy* (February 2015).

[28]DOD, *Strategic Sustainability Performance Plan FY 2016* (Sept. 7, 2016).

change in order to maintain an effective and efficient military. Mission planning and execution must include (1) identification and assessment of the effects of climate change on the DOD mission, (2) taking those effects into consideration when developing plans and implementing procedures, and (3) anticipating and managing any risks that develop as a result of climate change to build resilience.[29]

Congressional testimony from the Assistant Secretary of Defense for Energy, Installations, and Environment (Acting) has emphasized the need for the department to adapt to the impacts of climate change. The Assistant Secretary of Defense for Energy, Installations, and Environment has—since 2014—served as both DOD's Chief Sustainability Officer and the department's primary climate change adaptation official.[30] Since 2014, the Assistant Secretary of Defense (Acting) has testified annually that climate change is a top priority issue for DOD, outlining steps that the department is taking to mitigate the risk it poses. For instance, in March 2016 testimony before the House Appropriations Committee, the Assistant Secretary of Defense (Acting) stated that DOD would continue its efforts to develop the science and tools needed to meet the department's obligations to assess and adapt to climate change. The Assistant Secretary (Acting) reiterated that resilience to climate change continues to be a priority for DOD, explaining that—even without knowing precisely how or when the climate will change—the department knows it must build resilience into its policies, programs, and operations in a thoughtful and cost-effective way. One example the Assistant Secretary (Acting) provided is that sea level is rising and many coastal areas are

---

[29]DOD Directive 4715.21 was issued in accordance with the direction in Executive Order 13653. On March 28, 2017, the Presidential Executive Order on Promoting Energy Independence and Economic Growth (Executive Order 13783) rescinded Executive Order 13653. This rescinded executive order stipulated that, among other things, each agency was to develop or continue to develop, implement, and update comprehensive adaptation plans that integrate consideration of climate change into agency operations and overall mission objectives. According to an official from the Office of the Secretary of Defense, as of September 2017, DOD was working to determine the course of action the department will take with regard to its directive to comply with Executive Order 13783.

[30]In 2014 and 2015, the title of this position was the Deputy Under Secretary Of Defense (Installations and Environment). In 2016, the title of this position changed to the Assistant Secretary of Defense (Energy, Installations, and Environment).

subsiding or sinking, which impacts the operation and maintenance of DOD's existing installations and infrastructure.[31]

Funding Sources to Pay for U.S.-Funded Infrastructure Overseas

The military services use several funding sources for the design and construction of infrastructure overseas. Projects that cost over $1 million are generally funded through military construction appropriations. The construction cost estimates are prepared for the planning, design, and construction phases of a construction project.[32] DOD is authorized to use available operations and maintenance appropriations for unspecified minor military construction projects of $1 million or less,[33] and in its fiscal year 2017 budget, the department also planned to allot operations and maintenance funding to the maintenance of equipment.[34] Further, DOD's facilities sustainment, restoration, and modernization program provides funds for installation-level efforts to maintain, improve, and adapt existing facilities to meet current or new conditions and standards.

---

[31]*Installations, Environment, Energy and BRAC: Hearing Before the Subcomm. on Military Construction, Veterans Affairs and Related Agencies of the H. Comm. on Appropriations*, 114th Cong. (2016) (statement of Pete Potochney, performing the duties of Assistant Secretary of Defense (Energy, Installations and Environment)).

[32]DOD, Unified Facilities Criteria 3-740-05, *Handbook: Construction Cost Estimating* (Nov. 8, 2010).

[33]10 U.S.C. § 2805(c). For more information on statutory authorities for carrying out military construction projects, see appendix II of GAO's report, *Defense Infrastructure: Actions Needed to Enhance Oversight of Construction Projects Supporting Military Contingency Operations*, GAO-16-406 (Washington, D.C.: Sept. 8, 2016).

[34]Office of the Under Secretary of Defense (Comptroller)/Chief Financial Officer, *Operation and Maintenance Overview Fiscal Year 2017 Budget Estimates* (February 2016).

DOD Collaboration with Host Nations on Infrastructure Cost Sharing

DOD engages in cost-sharing activities with some host nations on infrastructure projects at overseas locations.[35] Specifically, host nations can provide financial support by directly or indirectly sharing installation and operational costs with DOD.[36] According to DOD's Facilities Investment and Management Office, it is DOD policy to actively seek host nation cost-sharing support from countries hosting U.S. forces to help cover U.S. construction requirements, when possible. Cost-sharing activities vary from country to country and are typically implemented by a variety of bilateral agreements.[37]

Officials from DOD's Facilities Investment and Management Office told us that DOD currently collaborates with Kuwait, Germany, Japan, and Korea on national cost-sharing activities. The Government of Kuwait has provided funds for construction projects mutually beneficial to U.S. and Kuwaiti military forces. According to a DOD official, since May 2016, Congress has received notification of eleven infrastructure projects that will be funded with approximately $163 million in cash contributions from the Government of Kuwait. Also, according to that same official, DOD recently approved a payment-in-kind project in Germany. In this type of project, credit provided by a host nation allows DOD to build, repair, or modernize its facilities in the host country. One of the chief methods of

---

[35] According to the Office of the Assistant Secretary of Defense for Energy, Installations, and Environment, "cost-sharing activities" refers to host-nation funding and support for overseas infrastructure used by DOD and includes host nation funded construction programs and projects, realignment and relocation efforts, in-kind and residual value payments, logistics, labor, and utilities cost-sharing, and any other installation or operational costs shared between the United States and host nations either directly or indirectly.

[36] Direct cost sharing refers to categories of support for stationed U.S. forces budgeted for by a host nation and includes, but is not limited to, costs borne by host nations in support of stationed U.S. forces, for rents on privately owned land and facilities, facility improvements, labor, and utilities. Indirect cost sharing covers cost deferrals and waivers for U.S. forces stationed in a host nation and includes reduced or waived rents on government-owned land and facilities used by U.S. forces, tax concessions, and waived customs duties.

[37] For the purposes of this report, agreements refers to a multilateral or bilateral agreement, such as a base rights or access agreement, a Status of Forces Agreement, a Special Measures Agreement, a Memorandum of Understanding or Agreement, a Technical Arrangement, a Local Implementing Agreement (that is within the scope of the umbrella or master agreement), or any other instrument defined as a binding international agreement in accordance with DOD Directive 4715.05, *Environmental Compliance at Installations Outside the United States* (Nov. 1, 2013) and DOD Directive 5530.3, *International Agreements* (June 11, 1987) (incorporating change 1, Feb. 18, 1991) (certified current as of Nov. 21, 2003).

DOD's collaboration with host nations on national cost-sharing activities is through host-nation-funded construction programs, in countries such as Japan and Korea. These host-nation-funded construction programs provide significant financial support to DOD for realignment and relocation efforts—for example, moving U.S. forces from one installation to another in a host nation—and infrastructure improvement programs. For example, the Government of Japan has spent, as of March 2017, over $23 billion through the Japan Facilities Improvement Program and, as of November 2016, approximately $17 billion through the Defense Policy Review Initiative for the construction of DOD facilities in Japan.[38] Similarly, a DOD installation-level official told us that U.S. Forces Korea receives roughly $350 million annually in support of large-scale construction projects from the Republic of Korea.[39]

Cost-sharing activities typically consist of three key collaborative mechanisms: national agreements, host-nation building standards, and negotiation processes. According to DOD officials, certain national agreements between the United States and host-nation governments implement cost-sharing activities, such as host-nation-funded construction programs. For example, a Special Measures Agreement establishes annual funding contributions for host-nation-funded construction programs between the United States and the Republic of Korea. The Japan Facilities Improvement Program also establishes funding contributions for host-nation funded construction between the United States and the Government of Japan. Infrastructure projects funded through cost-sharing activities are usually designed and built

---

[38]Host-nation-funded construction programs in Japan include the Japan Facilities Improvement Program, the Defense Policy Review Initiative, the Special Action Committee on Okinawa, and the Facilities Adjustment Panel. An example of a host nation funded construction project in Japan is the reconstruction of a 200-meter seawall at an ammunition depot. This $2.8 million project will be funded through the Japan Facilities Improvement Program.

[39]Host-nation-funded construction programs in Korea include the Republic of Korea Funded Construction program, the Yongsan Relocation Plan, and the Land Partnership Plan. According to installation-level officials, examples of planned host-nation funded construction projects in Korea include the construction of a consolidated communications facility, a special operations command facility, a training swimming pool, and an upgrade to pier operations. These projects will be funded with Republic of Korea in-kind funds. DOD does not consider the Yongsan Relocation Plan and the Republic of Korea portion of the Land Partnership Plan to be host-nation support, as the funds received from the Republic of Korea support the host nation's requests. However, we are including the Yongsan Relocation Plan and Land Partnership Plan in this report because the resources are provided by the host nation.

using a combination of host-nation building standards and DOD's Unified Facilities Criteria.[40] These criteria are administered and authorized by the military services, in consultation with the Office of the Assistant Secretary of Defense for Energy, Installations, and Environment, which are responsible for the Unified Facilities Criteria program. According to DOD officials, typically, projects are built to standards equivalent to the Unified Facilities Criteria except when host-nation building standards are more stringent. Depending on the host nation and combatant command, reconciling differences between the Unified Facilities Criteria and host-nation building standards can require close collaboration between the military services, the sub-unified commands, and host-nation officials. Once host nations have completed construction for infrastructure funded through these cost-sharing activities, DOD provides funds and resources for the facilities' sustainment and maintenance. According to the Office of the Under Secretary of Defense for Acquisition, Technology, and Logistics, often DOD uses this type of infrastructure for 50 or more years.

Prior to negotiations with host nations, DOD's Office of the Assistant Secretary of Defense for Strategy, Plans and Capabilities works with the combatant commands, the sub-unified commands, and the Secretaries of the military services and the Commandant of the Marine Corps to develop a list of key negotiation topics. According to officials from this office and State's Office of Security Negotiations and Agreements, State works with DOD to refine this list of key negotiation topics and generally conducts negotiations with host nations on DOD's behalf. The Circular 175 Procedure, a set of regulations developed by State, establishes requirements for negotiating and concluding international agreements.[41] According to State, for "significant" international agreements, DOD must submit a request for authorization to negotiate, conclude, or terminate an international agreement. In our previous work on host-nation support negotiation, we found that generally DOD participates in the Circular 175 process by documenting a proposed negotiating strategy in a memorandum and submitting it to State for review and approval.

---

[40]According to DOD, the Unified Facilities Criteria program unifies all technical criteria and standards pertaining to planning, design, construction, and operation and maintenance of real property facilities.

[41]Circular 175 sets out the process by which the Secretary of State or his or her designee authorizes negotiations and conclusions of treaties and other international agreements. The objectives of the Circular 175 process include ensuring that the making of international agreements is carried out within constitutional and other appropriate limits and with the required involvement by the Department of State.

DOD Installations Have Identified Operational and Budgetary Risks Posed by the Impacts of Weather Effects Associated with Climate Change, but Do Not Consistently Track the Costs of These Impacts

DOD has reported that the potential impacts of a changing climate represent risks and DOD overseas installations have identified operational and budgetary risks posed by the impacts of weather effects associated with climate change. However, the installations do not consistently track the costs of these impacts. As a result, the military services lack the information they need to adapt infrastructure at overseas installations to weather effects associated with climate change and develop accurate budget estimates for infrastructure sustainment.

DOD Installations Have Identified Impacts and Risk to Training, Testing, and Mission Operations Posed by Weather Effects Associated with Climate Change

DOD has reported that the potential impacts of a changing climate, whether in the near term or in the future, represent risks to the department.[42] Specifically, according to DOD's key strategy documents and policy on climate change, potential and observed climate change impacts represent risks to the department's infrastructure and operations. At a majority of the 45 installations we visited or contacted, officials provided examples of how the impacts of weather effects associated with climate change pose operational risks to training, testing, or mission operations.

Training and Testing Operations

DOD manages both built and natural infrastructure to support the military services' training and testing activities. In its 2014 Roadmap, DOD stated that the department must be able to train its forces to meet the evolving nature of the operational environment and that climate change will have serious implications for DOD's ability to ensure military readiness in the future. During our review, installation officials identified multiple examples of how weather effects associated with climate change had impacted training and testing locations. Installation officials also discussed the potential for additional risks in the future for several of these locations.

---

[42]Survey Summary Report.

- At a missile testing range in the Pacific, officials stated that in 2008, severe tides caused flooding at two antenna facilities, damaging air conditioning equipment and causing communication cables to corrode faster than normal. In addition, a 2015 storm degraded pier operations, disrupted transportation schedules, and required the diversion of equipment to assist in the recovery of four boats that sank. The storm also damaged piers, roofs, shorelines, and beaches, and caused multiple sinkholes to develop. That same year, severe winds caused a delay in evaluating the operation of certain equipment used in testing. Also, in 2016, more severe winds delayed a missile assembly test.

- At a DOD facility in the Middle East, officials stated that the number of black flag days—when non-mission essential physical training and strenuous exercise must be suspended for all personnel—had increased because the facility was experiencing more days of extended heat.[43] As a result of extreme temperatures, outdoor training was been suspended.

- At a DOD training range in the Pacific, according to our review of installation documentation, coastal erosion potentially linked to sea level rise threatens to increase flooding and the loss of both U.S. and coalition missions at the training area. According to installation officials, the training range has the mission of providing highly realistic training for service, joint, and coalition combat aircrews. Specifically, the range supports approximately 1,400 training sorties per year for U.S. and coalition forces and the erosion has the potential to degrade the readiness of units training at the range. According to officials, installation civil engineers stopped the rapidly advancing erosion about 1 week before the erosion would have destroyed a tower that contains essential range equipment and serves a range safety function.

Mission Operations

During our review, installation officials identified multiple examples of how weather effects associated with climate change had already impacted mission operations. Moreover, they noted that the potential exists for additional impacts at these locations.

---

[43]On black flag days, installations experience high temperature. For example, this DOD installation reported that on black flag days, heat stress will occur in most cases, heat stroke is likely with continued exposure, and outdoor work should be limited to critical missions only (requiring a commander's approval).

- At a DOD installation in Europe, according to DOD documentation and installation officials, flooding has increased in frequency over the last 12 years and is a significant threat to the base's infrastructure and mission. During past flooding events, the installation's runways were flooded, preventing the installation from launching or recovering aircraft, often for several days at a time. Based on our review of documentation provided by installation officials, the operational risks posed by flooding at this DOD installation are potentially significant, given that the aircraft of one unit operating out of the installation provide real-time intelligence, surveillance, reconnaissance, and ground attack in support of counter-terrorism plans. Aircraft from another of the installation's units deliver globally integrated, persistent intelligence, surveillance, and reconnaissance in support of national security objectives. Figure 1 shows flooding at the installation.

Figure 1: Flooding on Roads and in Facilities (Left) and Mud Covering the Ramp of a Runway (Right) at a Department of Defense Installation in Europe

Source: Department of Defense. | GAO-18-206

- At a DOD installation in Europe, officials stated that sea level rise has the potential to cause flooding that would negatively impact a variety of infrastructure essential to supporting several military services' missions. The at-risk infrastructure sits close to a canal connected to the sea, and includes DOD prepositioned stocks supporting multiple areas of operations, a logistics transportation hub, storage facilities that house ammunition, and docks used for loading and unloading the ammunition from barges on the canal. For example, a substantial portion of the ammunition storage area is either below the current mean sea level or just above mean sea level.[44] Because of this—according to installation officials—that portion of the ammunition storage area is potentially vulnerable to canal flooding caused by high water events such as mean sea level rise coupled with high tides or storm surge. In addition, officials explained that the warehouses and facilities that support the prepositioned stock mission—which includes the maintenance of thousands of wheeled vehicles—is also potentially vulnerable to flooding from the canal. This area of the installation is made more vulnerable by a storm water drainage system that officials characterized as "failing." Figure 2 shows the canal's water level in relation to the docks.

Figure 2: Water Level of a Canal Adjacent to Ammunition Loading Docks at a Department of Defense Installation in Europe

Source: Department of Defense. | GAO-18-206

- At a DOD installation in the Pacific, officials explained that access to a munitions complex is threatened by erosion and flooding caused by increasingly frequent and intense rain events. For example—according to installation officials—during heavy rain events, both access points to the munitions complex can be inaccessible. Also, in the past, erosion has led to a significant landslide that limited access to the munitions complex. These threats to access represent potentially serious operational impacts given that the munitions

[44]According to a multiagency report, mean sea level can be defined as an average sea level over a specified time, such as annual or monthly mean sea level. Hall, J.A., S. Gill, J. Obeysekera, W. Sweet, K. Knuuti, and J. Marburger *Regional Sea Level Scenarios for Coastal Risk Management: Managing the Uncertainty of Future Sea Level Change and Extreme Water Levels for Department of Defense Coastal Sites Worldwide* (April 2016).

complex is used to store munitions for multiple military services and a coalition partner and to resupply munitions to U.S. forces. In addition, installation officials stated that rains from a 2014 typhoon caused flash flooding, almost drowning two DOD personnel trapped at a security checkpoint. Figure 3 shows flooding at the checkpoint in 2014.

Figure 3: Security Checkpoint in Non-flood Conditions (Left) and During Flooding in 2014 (Right) at a Department of Defense Installation in the Pacific

Source: GAO and Department of Defense. | GAO-18-206

- At a DOD installation in the Pacific, officials stated that significant erosion at a pier limits the amount of traffic that can travel down the pier. The officials explained that the edges of the pier have begun to erode away, which has raised safety concerns, and the installation has been forced to use only the central portion of the pier.

- At a DOD installation in Africa, officials stated that high tides flood an area used for loading combat aircraft, and the resulting erosion also threatens the base's security fence, potentially compromising anti-terrorism and force protection. Specifically, if the security fence failed, it would temporarily compromise anti-terrorism/force protection standards and security personnel would need to guard the portion of the failing fence until repairs were made. Figure 4 shows the flooding and erosion that have occurred.

Figure 4: Flooding (Left) and Erosion (Right) at a Department of Defense Installation in Africa

Source: Department of Defense. | GAO-18-206

- At a DOD installation in the Caribbean, officials reported that a storm caused roof damage to a hospital. This severe weather resulted in operational impacts for several days by limiting transportation on installation roads and the shutdown of base operations.

The Impacts of Weather Effects Associated with Climate Change on Overseas Infrastructure Pose Budgetary Risks to the Military Services, but Costs Are Not Consistently Tracked

As discussed previously, DOD's key strategy documents and policy on climate change indicate that potential and observed climate change impacts represent risks to the department's infrastructure and operations. At many of the installations we visited or contacted, officials provided examples of both incremental and one-time costs associated with climate change impacts. Taken as a whole, these costs constitute budgetary risks that make DOD more vulnerable to fiscal exposure associated with the impacts of climate change. Officials from overseas locations in our sample were able to provide examples of some of the costs associated with climate change impacts. However, officials from a majority of locations we visited or contacted told us that they do not systematically track these costs and so could not provide the complete picture of the budgetary risks posed by climate change for infrastructure on their installations.

Incremental Costs

- At a DOD installation in Europe, officials stated that an increasing number of extreme wind storms are damaging equipment—such as helicopters (see fig. 5)—and infrastructure, requiring more frequent repair and resulting in increases in associated costs. Installation officials told us that they conducted numerous storm-related repairs from 2005 through 2016. Officials reported various costs associated with these repairs, which totaled almost $7 million. The officials also explained that there are incremental costs associated with the time that personnel spend recovering from extreme weather events, including the costs of personnel not performing their routine maintenance tasks. At this DOD installation in Europe, personnel spent 6 to 12 months to complete repairs following a storm. However, officials could not quantify these costs.

Figure 5: Helicopter Damaged by an Extreme Wind Storm at a Department of Defense Installation in Europe

A 2006 storm required at least $3.5 million in facility repairs and severely damaged this helicopter at this installation. In previous work, we found that while it is not possible to link any individual weather event to climate change, the impacts of extreme weather events provide insight into the potential climate-related vulnerabilities faced by the Department of Defense.

Source: Photo by Kent Harris / © 2006 Stars and Stripes. | GAO-18-206

- At several installations in Europe and the Pacific, officials stated that personnel must switch from their normal focus—supporting the operations and maintenance needs of the base—to focus instead on preparing for approaching extreme weather events, such as typhoons, and addressing the impacts of extreme weather events. In one example, officials from a DOD installation in the Pacific stated that their base typically experiences at least five typhoons annually and that they have noticed storms have been more severe over the past few years. Emergency staff work overtime at this installation, preparing sandbags and moving equipment indoors before a typhoon, and also after the storm has passed, examining sewer and power lines for damage and cleaning up debris. Officials explained that, to perform these storm-related tasks, installation personnel are switching from their typical focus—supporting the operations and maintenance needs of the base—to focus instead on preparing or cleaning up after storms. The officials stated that 50 people typically work to clean up after each typhoon, but the officials generally do not quantify all storm-related costs.

One-Time Costs

- At a DOD installation in Europe, officials stated that isolated incidents such as high winds in 1996 and an extreme rain event in 2001 caused severe damage totaling approximately $14.4 million. Also in Europe, officials stated that another DOD installation was impacted by a storm surge that, in 2001, damaged a breakwater of a fuel offloading facility,[45] requiring $50 million to repair.

- At installations we visited in the Pacific, officials also provided examples of one-time costs associated with climate change impacts. At another DOD installation, officials stated that heavy rain in 2011 caused a river to overflow, impacting safety, security, and mission operations. The repair costs were almost $10 million. In another example, significant erosion from rain events has limited road access to a DOD training range and weakened the supports for one third of the range's targets. Officials explained that further erosion will likely cause the target supports to collapse within 5 years. These officials also stated that a total of $24 million has been spent on erosion mitigation, with $1.5 million spent in 2014. Figure 6 shows the erosion at this training range.

---

[45]According to U.S. Army engineering documentation, a breakwater is—generally speaking—a shore-paralleling structure that reduces the amount of wave energy reaching the protected area by dissipating, reflecting, or diffracting incoming waves.

Figure 6: Erosion along Roads at a Department of Defense Training Range in the Pacific

Source: Department of Defense. | GAO-18-206

- Installations in other combatant commands' areas of responsibility also provided examples of costs associated with extreme weather events. For instance, officials reported that a 2016 hurricane caused substantial damage to infrastructure that supports a DOD undersea testing range in the West Indies. Examples include that over half of buildings have roof, siding, or door damage; employees were living in housing with tarps and sandbag roof patches; and the installation's ships sustained about $1 million of damage. The officials also stated that a variety of testing infrastructure is damaged and the testing range was only "minimally operational" at the time of our review. Officials further explained that seasonal wind and rain may result in the installation becoming non-mission capable for any or all missions. Repairs of this damage represent a large fiscal exposure for the service, estimated at $63 million. Also, at a DOD support facility in the Middle East, a 2013 storm event caused flooding and resulting damage to buildings, roads, utilities, and infrastructure used for training, with $1.2 million required for repairs.

According to Executive Order 13693, the head of each agency shall ensure that agency operations and facilities prepare for the impacts of climate change by, among other actions, calculating the potential cost and risk to mission associated with agency operations.[46] In addition, the

---

[46]Executive Order No. 13693, *Planning for Federal Sustainability in the Next Decade*, 80 Fed. Reg. 15869 (Mar. 19, 2015).

2014 *Roadmap* notes that the impacts of climate change may increase costs associated with facilities and require adaptations to how DOD plans and executes operations. The 2014 *Roadmap* states that DOD will—as it adapts to changing climate conditions—review and, as needed, modify models for facility maintenance and repair costs. Specific examples include DOD's guidance to components on how they prepare budget submissions, and—as the department adapts to changing climate conditions—the models the department uses for facility maintenance and repair costs.[47] Further, *Standards for Internal Control in the Federal Government* maintain that an agency's managers should use quality information to achieve the agency's objectives.[48]

As discussed above, officials from some installations in our sample were able to provide examples of costs associated with extreme weather or climate change impacts. According to our discussions with these officials, the costs often associated with climate change impacts are for the sustainment, restoration, or maintenance of installation infrastructure. For instance, installation officials provided examples of infrastructure that they anticipate will be more expensive to maintain due to the impacts of extreme weather or climate change. These include increasing rates of erosion caused by heavier rains at a DOD installation in the Pacific, requiring additional funding to maintain training areas, and air conditioning systems at a DOD installation in Europe that will need to be replaced more frequently due to longer periods of extreme temperatures.

Also, officials from the military services' overseas regional organizations and installations in our sample explained that installations generally have the capability to track these types of costs associated with extreme weather and climate change. For instance, according to installation and

---

[47] According to our previous work on DOD's funding of installations' sustainment and maintenance requirements, DOD has models that predict future requirements based on known inputs specific to a facility's category type, service, and location. Each year, funds are allocated to installations for sustainment activities, and in most cases that funding is based on a percentage of the facilities sustainment model's estimated requirements for each installation. Installations also receive additional sustainment funding during the fiscal year, when available. See GAO, *Defense Facility Condition: Revised Guidance Needed to Improve Oversight of Assessments and Ratings*, GAO-16-662 (Washington, D.C.: June 23, 2016) and *Defense Infrastructure: Continued Management Attention Is Needed to Support Installation Facilities and Operations*, GAO-08-502 (Washington, D.C.: Apr. 24, 2008).

[48] GAO, *Standards for Internal Control in the Federal Government*, GAO-14-704G (Washington, D.C.: Sept. 10, 2014).

regional officials, both Air Force and Marine Corps headquarters can use a special accounting code to track funding provided to installations to recover from the damage caused by extreme weather events. In addition, officials from installations we visited or contacted provided examples of how they tracked costs for infrastructure repair related to climate change impacts. For example, a DOD installation in the Pacific had tracked costs over multiple years for needed repairs due to typhoon damage, and another installation in the Pacific tracks certain costs associated with preparation before storm events.

Consistently collecting cost data at the installation level would help DOD meet the direction outlined in Executive Order 13693, the 2014 *Roadmap*, and *Standards for Internal Control in the Federal Government*. According to officials from the Office of the Secretary of Defense, their expectation is that within the next 5 years, DOD will be using data on actual facility repair costs to improve the accuracy of facility maintenance and repair costs.[49] For example, according to DOD's 2016 *Sustainability Plan*, in fiscal year 2015 the Army began an evaluation of the impact of severe weather events on its training lands and infrastructure. As part of this evaluation, the analysis of repair cost data and recovery times will enable the Army to—among other things—make appropriate climate-related adjustments when developing future budgets. However, the majority of installations in our sample indicated that they did not consistently track costs associated with climate change impacts because, according to an official from the Office of the Secretary of Defense and several installations' officials, there is no requirement from the Office of the Secretary of Defense or the military services for them to do so. Without a requirement for installations to track costs associated with climate change in a systematic way, the military services will lack the information they need to adapt infrastructure at overseas installations to weather effects associated with climate change and develop accurate budget estimates.

---

[49]Currently, the services are to assess the condition of buildings, pavement, and railway facilities using Sustainment Management System software tools developed by the U.S. Army Corps of Engineers. For other types of facilities that cannot be assessed with these software tools, the services are to determine existing physical deficiencies and estimate the cost of maintenance and repairs using established industry cost guides. See GAO-16-662.

DOD Has Collected Incomplete Data on Overseas Climate Impacts, Which Creates Challenges for Effectively Managing Climate Risks

DOD has taken steps to obtain information to help manage the operational and budgetary risks of weather effects associated with climate change, but the approach used to gather survey data resulted in the department collecting information that was incomplete and not comprehensive. First, DOD exempted certain sites from the total set of potential sites to be surveyed. Second, DOD did not consider all locations from a list of key national security sites when creating the total set of potential sites to be surveyed. As a result, DOD's survey did not obtain information related to the operational and budgetary risks posed by weather effects associated with climate change at a variety of its overseas sites, including a significant portion of the department's most important overseas sites. This is information that is critical for DOD's plans to manage climate change risk.

As discussed previously, DOD has reported that the potential impacts of a changing climate, whether in the near term or in the future, represent additional risks that the department must incorporate into its planning and risk management processes. Further, in key documents that establish policy and goals regarding climate change adaptation, DOD has stressed the role that risk management plays in the department's approach to climate change adaptation. For example, according to the 2010 *Quadrennial Defense Review*, the department will regularly reevaluate climate change risks in order to develop policies and plans to manage the impacts posed by climate change on DOD's operating environment, missions, and facilities. DOD reiterates this theme in its 2014 *Quadrennial Defense Review*, stating that the department will continue to seek mitigation of the risks posed by climate change impacts. Also, in the 2014 Roadmap and in its Survey Summary Report on the screening level vulnerability assessment, DOD stated that it would use assessments from several surveys to, among other things, gather data on installations' vulnerability to operational and budgetary risks posed by climate change and manage these risks. Further, DOD's 2016 *Sustainability Plan* states that the department's sustainable enterprise—including climate change adaptation—requires resilience in the face of an unknown future operating environment and a robust risk management framework to address changes when they occur.

According to our previous work on how governments can best use information to manage the risks posed by climate change, reducing these risks require making decisions based on reliable and appropriate information about, among other things, the past and future climate.[50] In additional previous work on assessment of surveys, we have outlined generally accepted survey research principles, derived in part from U.S. Office of Management and Budget guidelines.[51] Specifically, leading practices for the implementation of surveys include, among other practices, ensuring the quality of survey data. Also, as previously discussed, *Standards for Internal Control in the Federal Government* maintain that an agency's managers should use quality information to achieve the agency's objectives.[52]

According to DOD's Survey Summary Report on its Screening Level Vulnerability Assessment Survey process, the survey is the first step in the department's ongoing process to manage the risks associated with a changing climate to the DOD mission, installations, and training ranges. To identify installations with climate-related vulnerabilities, the Office of the Secretary of Defense worked with the military services to conduct the survey at over 3,500 sites, including almost 400 overseas sites, from 2013 to 2015.[53] The survey sought to collect data on the seven extreme weather effects DOD has identified as being commonly associated with climate change—flooding due to storm surge, flooding due to non-storm surge events, extreme temperatures, wind, drought, wildfire, and changes in mean sea level. In our previous work examining the impacts of climate change on DOD infrastructure, we noted that it was DOD's goal for the military services to complete the survey as a comprehensive assessment of all installations, assessing the potential impacts of climate change on DOD's missions.[54]

---

[50]GAO, *Climate Information: A National System Could Help Federal, State, Local, and Private Sector Decision Makers Use Climate Information*, GAO-16-37 (Washington, D.C.: Nov. 23, 2015).

[51]GAO-11-702 and the Office of Management and Budget, Statistical Programs and Standards, *Standards and Guidelines for Statistical Surveys* (Washington, D.C.: September 2006).

[52]GAO-14-704G

[53]According to DOD, the survey was developed in concert with the military services, Defense Logistics Agency, Washington Headquarters Services, and the department's infrastructure-owning and -managing components. In this report, we focus on the military services' administration of the survey to overseas sites because the military services own or use the vast majority of overseas infrastructure used by DOD.

[54]GAO-14-446.

The Survey Summary Report and military service documentation on their post-survey plans provide several examples of the military services' plans for using survey data to manage the risks posed by the impacts of weather effects associated with climate change. According to the report, both Army and Navy installations plan to use their survey responses as they prepare future installation-specific plans. In addition, the Army should be able to prioritize potential future installation-level actions to meet mission requirements. The Air Force plans to incorporate survey data into existing climate metrics that are already part of its installation planning processes. Also, the Marine Corps plans to integrate survey data, along with other information, into its existing assessment and planning processes to manage mission risks.

We found that the military services have collected some data on climate changes impacts through this survey, but DOD's approach to obtaining survey data was both incomplete and not comprehensive. First, DOD exempted certain sites from the total set of potential sites to be surveyed. Specifically, we found that 73 of 198 (37 percent) of exempted sites in overseas locations were omitted from the survey without adequate justification.[55] Instructions issued as part of the survey's implementation stipulated that a site could be exempted from the survey if DOD officials decided that a certain site was not a high priority for screening purposes at the time of the survey.[56] Further, the instructions required a reason for any such determination. For these exempted sites, we determined that the explanations were insufficient because, while the survey instructions required a reason as to why a site was not deemed a high priority for screening, they did not provide complete information on why the sites' infrastructure—and the missions supported by that infrastructure—was

---

[55]The exempted sites were in the U.S. Northern, Southern, Central, European, and Pacific Commands.

[56]According to our review of DOD's Survey Summary Report, more than 3,500 sites completed surveys in the United States and overseas, while officials exempted about 2,250 additional sites from completing surveys. Another reason a site could be exempted was if the location was included in the survey response from another site. In our review, we did not review overseas sites that were exempted because they were included in another site's survey. Instead, we reviewed about 200 overseas sites that were exempted because DOD officials decided the sites were not a high priority for screening purposes at the time of the survey.

not vulnerable to the seven types of climate change impacts that the survey addressed.[57] Table 2 summarizes the results of our review.

Table 2: Number of Department of Defense (DOD) Overseas Locations Exempted from the Survey

Characterization of exemption	Number of exempted sites[a]	Percentage of total
Exempting officials provided a sufficient explanation for exemption	125	63
Exempting officials did not provide a sufficient explanation of why the site's infrastructure, and thus mission, is not vulnerable to the seven identified types of climate change impacts	73	37
Total	198	100

Source: GAO analysis of DOD data. | GAO-18-206

[a]A site could be exempted if the location was included in the survey response from another site. In our review, we did not review overseas sites that were exempted because they were included in another site's survey. Instead, we reviewed about 200 overseas sites that were exempted because DOD officials decided the sites were not a high priority for screening purposes at the time of the survey.

In these 73 cases, while officials provided an explanation for exempting a site, they did not fully explain why information on these sites' potential vulnerabilities should not be included as part of their military services' risk assessment efforts. For example:

- Officials exempted several communications sites in the Pacific because they were either "unoccupied" or reported to be "20 meters above sea level." However, occupancy of a site does not have any bearing on whether the site's infrastructure is vulnerable to the impacts of climate change. Further, exempting a site based on its invulnerability to one type of climate change impact (i.e., sea level rise) does not sufficiently explain whether the site is vulnerable to the six other types of impacts that the survey includes.

---

[57]As discussed above, through the survey, sites could report the following climate change impacts: drought, extreme temperatures (hot or cold), flooding and other impacts due to non-storm surge events, flooding due to storm surge, implications of increased mean sea level, wildfire, and wind.

- Several sites in Germany, Belgium, and Oman were exempted because, according to survey responses, the sites were not vulnerable to any of seven types of impacts included in the survey. For example, the reason provided for three sites was "no issues with climate change." However, the responses did not explain why the sites will not be impacted by any of the seven types of impacts. Given that the survey's purpose was to assess sites' vulnerability to these impacts, exempting sites without providing an explanation is inconsistent with the stated purpose of the survey.

Second, DOD did not consider all locations from a list of key national security sites when creating the total set of potential sites to be surveyed. Specifically, we found that the military services' data collection efforts were not comprehensive, namely because the survey was not administered to all of DOD's key overseas sites. The survey was administered to fewer than half of the sites found on DOD's list of these locations—called the Enduring Locations Master List (Master List).[58] According to DOD Instruction 3000.12, this list records sites that DOD maintains and operates from in foreign locations and U.S. territories in order to accommodate an adjustable force presence and the necessary flexibility to respond to crises and ensure homeland defense. Further, a location is enduring when DOD designates it for strategic access and uses it to support U.S. security interests for the foreseeable future.[59] Enduring locations are categorized as either major operating bases, forward operating sites, or cooperative security locations, each of which has a varying degree of DOD-owned or -managed infrastructure. There are examples of enduring locations—from several different military services—that received surveys and reported a variety of impacts from weather effects associated with climate change. This indicates that sites from the Master List, like other locations discussed in this report, are vulnerable to operational and budgetary risks associated with climate change.

According to our review of survey data and discussions with military service officials, the military services did not use the Master List as a source for identifying survey respondents. Instead, the military services used DOD's Real Property Asset Database as the only source for

---

[58]Because individual locations on the Master List are classified, we do not discuss them in this report.

[59]DOD Instruction 3000.12, *Management of U.S. Global Defense Posture (GDP)* (May 6, 2016).

surveyed locations. The Real Property Asset Database includes real property records for owned and leased assets directly managed by DOD components.[60] Officials from various offices within the Office of the Secretary of Defense, the military services, and the Joint Staff explained that they believed the Real Property Asset Database to be the most comprehensive list of real property in which DOD has equity and were not aware of the Master List. In addition, according to an official from the Office of the Secretary of Defense, the Master List may not be fully reconciled with sites recorded in the Real Property Asset Database. Further, military service officials stated that, generally, they did not send surveys to two types of enduring locations on the Master List—forward operating sites and cooperative security locations—because the officials believed that because those locations are sites where host nations owned infrastructure, the military services were not investing significant appropriated resources in defense infrastructure at these locations.[61] However, this is not the case. For example, according to our review of budget data for fiscal years 2015 through 2017, certain military services plan to spend more than $354 million on infrastructure projects at selected forward operating sites and cooperative security locations. Because over half of the enduring locations on DOD's Master List were excluded from the survey, DOD's survey did not obtain information related to the operational and budgetary risks posed by weather effects associated with climate change at a variety of its overseas sites, including a significant portion of its most strategically important overseas sites. This is information that is critical for DOD's plans to manage climate change risk. According to DOD officials, the department has acknowledged these challenges and is planning to determine the appropriate course of action to capture pertinent information in the future.

---

[60] According to our prior work on DOD's Real Property Asset Database, the database is the single authoritative source for all data on the department's real property inventory. See GAO, *Defense Infrastructure: More Accurate Data Would Allow DOD to Improve the Tracking, Management, and Security of Its Leased Facilities*, GAO-16-101 (Washington, D.C.: Mar. 15, 2016).

[61] Each of the three types of enduring locations may contain infrastructure in which the military services have invested resources. According DOD Instruction 3000.12, major operating bases are primarily characterized by the presence of permanently assigned U.S. forces and robust infrastructure. Forward operating sites are primarily characterized by the sustained presence of allocated U.S. forces, with infrastructure and quality of life amenities consistent with that presence, capable of providing forward-staging for operational missions and support to regional contingencies. Contingency security locations are characterized primarily by the periodic presence of allocated U.S. forces, with little or no permanent U.S. military presence or U.S.-controlled infrastructure, used for a range of missions and capable of supporting surge requirements for contingencies.

DOD Has Included Climate Change Adaptation in Infrastructure Guidance, but Integration into Installation-Level Planning Is Limited and Design Standards Have Not Been Updated

Approximately one-third of the installations in our sample had integrated climate change adaptation into their installation and natural resource plans but at these installations it was rare for climate change adaptation to be included in project designs. Some installation planning officials reported that they have taken steps to begin integrating climate change adaptation into planning efforts and projects as required, but the lack of key guidance, training, and updated design standards criteria that reflect climate change concerns hampers the planners' ability to consistently incorporate climate change adaptation into plans and individual projects.

DOD Requires Installation Plans to Include Climate Change Adaptation, but Installation Planners' Integration Efforts Are Limited

In its 2010 and 2014 *Quadrennial Defense Reviews*, DOD noted the importance of developing plans to adapt installations and facilities to the impacts of climate change. Also, in its 2014 *Roadmap*, the department emphasized the importance of integrating climate change adaptation into installation planning efforts.[62] DOD guidance has also instructed installation-level planners to integrate climate change adaptation into Installation Master Plans (master plans). Specifically, DOD's Unified Facilities Criteria for *Installation Master Planning* states that installation planners can prepare a master plan that sustainably accommodates future change by incorporating current needs and mission requirements into a vision with clear goals and measurable objectives.[63] The guidance further states that the military services' master planners are to understand, monitor, and adapt to—among other things—changing climatic conditions. Additionally, DOD Instruction 4715.03 discusses DOD's natural resources conservation program. Although this instruction does not apply to DOD sites outside the United States, based on our review of documentation provided by officials from overseas' locations, these officials follow the procedures in the instruction. Further, according

---

[62]As of January 2016, DOD planned to integrate climate change considerations into 43 separate pieces of DOD policy and guidance that address installation management, basing, or project design.

[63]DOD, Unified Facilities Criteria 2-100-01.

to discussions with officials from the Office of the Secretary of Defense and military services' headquarters, they expect overseas installations to be following the instruction's sections that address climate change. The instruction states that all DOD natural resources conservation programs shall be integrated with installation planning and programming. The instruction further states that for Integrated Natural Resources Management Plans (natural resources plans), all DOD components are to utilize existing tools to assess the potential impacts of climate change to natural resources on DOD installations, to the extent practicable and using the best science available.[64] Further, according to DOD's 2016 *Sustainability Plan* and Survey Summary Report, each of the military departments is planning to integrate climate change adaptation into installation-level plans, such as master and natural resources plans, as a way to address the risks posed by climate change impacts.

The importance of integrating climate change adaptation into master planning is demonstrated by the threat to installations posed by flooding. In its 2014 memorandum on installation floodplain management, DOD recognizes that due to changes in climate—as well as near term weather variability—it is imperative that the department plan and manage those facilities vulnerable to flooding to ensure the resilience of the installations and facilities required to support its missions.[65] Officials from installations we visited or contacted provided examples that illustrate both the benefit of integrating climate change adaptation into installation-level planning and the risks of not doing so. For example:

- DOD officials reported that at one of DOD's Caribbean installations, the military service is making repairs to a mission-critical wharf located within the 100-year floodplain.[66] According to DOD documentation, these repairs are intended to make the wharf more resilient to future changes in the climate.

---

[64] DOD Instruction 4715.03.

[65] Office of the Under Secretary of Defense for Acquisition, Technology, and Logistics Memorandum, *Floodplain Management on Department of Defense Installations*, (Feb. 11, 2014).

[66] According to DOD's 2014 memorandum, the United States has experienced significant storms in recent years. These events impacted DOD installations, causing—among other things—damage due to flooding and storm surges. The most substantial damage has occurred in areas where, for mission reasons, DOD facilities are located within the areas designated by the Federal Emergency Management Agency as those within the 1 percent annual chance of flood boundary. This was formerly known as the 100-year floodplain hazard area.

- At a DOD installation in Europe, rapid erosion caused by flooding is threatening a communications system that supports a broad set of missions. The erosion is occurring on the side of the hill on which three satellite dishes sit and on which a large gully is opening due to erosion (see fig. 7). Officials estimate that, given the rapid pace of erosion, the gully will reach the base of the first satellite dish in 1 to 2 years and the dish will likely "fall off the face of the hill." According to officials, replacing this satellite dish would cost approximately $70 million. While the most recent version of this installation's master plan does address climate change, the site was built before the most recent version of the master plan.

Figure 7: Hillside Impacted by Erosion at a Department of Defense Installation in Europe

Source: GAO. | GAO-18-206

Installation officials also provided examples of the importance of integrating climate change adaptation into natural resources planning, highlighting how species management can affect the availability of training areas. DOD's 2015 guidance on sustaining access to training areas instructs planners to evaluate the risks to training and range capability from the impacts of climate change trends, including impacts to threatened and endangered species and species at risk.[67] Also, DOD's 2016 *Sustainability Plan* states that changing temperature and precipitation patterns accompanying climate change may cause shifts in the composition or geographic range of some species. For instance, DOD may face the need to set aside more land and use personnel hours for the management of endangered or threatened species. In our previous work on DOD adaptation to climate change impacts, we noted how these impacts may have caused a protected turtle species to nest on a section of beach where it previously had not nested, making that section unavailable for training activities at the time of the turtles' nesting.[68] Illustrating this type of impact, officials from a DOD installation in the Pacific explained that they dedicated time to relocating the eggs of sea turtles from a section of the beach vulnerable to increasing storm surge (see fig. 8).

Figure 8: Movement of Sea Turtles across a Department of Defense (DOD) Beach, Indicated by Arrows (Left) and Sea Turtle Hatchlings (Right), at a DOD Installation in the Pacific

Source: DOD. | GAO-18-206

[67]DOD Instruction 3200.21, *Sustaining Access to the Live Training Domain* (Sept. 15, 2015) (incorporating change 1, effective Nov. 4, 2015).

[68]GAO-14-446.

However, we found that the majority of existing installation-level plans for overseas locations that we reviewed do not include climate change adaptation. Specifically, of the 50 master plans issued in or after January 2013 that we reviewed, only 19 addressed climate change. Moreover, of the 27 natural resources plans issued in or after January 2012 that we reviewed, only 9 addressed climate change.[69] According to military service officials from the regional and installation level, planners' integration of climate change adaptation into overseas installation-level plans is limited mainly because planners were often unaware of the master planning requirement to integrate climate change adaptation into master plans and the military services' goal to integrate climate change adaptation into natural resources plans.

The military services have established processes used to train planners on how to develop master plans and natural resources plans, but the training lacks direction on how to integrate climate change adaptation. DOD has taken some steps to facilitate the integration of climate change into these types of plans, but installation officials explained that the military services' headquarters organizations did not consistently communicate the department-wide requirement and goal to integrate climate change adaptation into master plans and natural resources plans, or provide direction or training on how integration should occur. For example, DOD provided updated information through its manual on natural resources plan development in 2013 to include suggestions on how planners could take steps to address climate change in their installations' natural resources plans.[70] However, at the military service level, training has been inconsistent as illustrated by the following examples.

---

[69]For the purposes of this review, we analyzed installation master plans released in or after January 2013, after the issuance of DOD Unified Facilities Criteria 2-100-01, *Installation Master Planning Standard*, in 2012. In addition, we analyzed natural resources plans released in or after January 2012, after the issuance of DOD Instruction 4715.03, *Natural Resources Conservation Program*, in 2011.

[70]DOD Manual 4715.03, *Integrated Natural Resources Management Plan (INRMP) Implementation Manual* (Nov. 25, 2013). Similar to the instruction, DOD Manual 4715.03 does not apply to overseas installations. However, according to DOD officials, the manual's guidance is used by overseas installations in preparation of their natural resource plans.

- An Air Force instruction on comprehensive planning mentions "climatic vulnerability," but does not provide specific instruction on integrating climate change adaptation into these plans.[71]
- Officials at a DOD installation in the Pacific provided a list of required trainings for installation-level planners to develop installation master plans and natural resources plans. The installation officials specified that—based on their understanding of guidance—the installation is not required to plan for climate change and thus these provided trainings do not address climate change.
- Officials at a DOD installation in the Pacific also identified required trainings for installation plan development, which include multilevel workforce development courses, online practicums, as well as online training resources offered by the U.S. Army Corps of Engineers. Based on our review of these materials, the trainings do not provide instruction on how to integrate climate change adaptation into installation plans.
- Officials at a DOD installation in the Pacific stated that installation planners underwent a 3-day natural resources plan development training that discussed, among other topics, climate change. Although the training directed installation planners to integrate climate change into the natural resources plans, the training did not provide instruction on how to do so.

Also, officials from U.S. European Command and U.S. Africa Command told us that they were not aware of the requirement to integrate climate change adaptation into master or natural resource plans.

If installation planners are not aware of the requirement that climate change adaptation is to be integrated into installation-level planning efforts and do not receive adequate training on how to do this, then DOD's overseas installations will likely have inadequate plans that do not include adaptation measures.

---

[71]Air Force Instruction 32-7062, *Comprehensive Planning* (Dec. 18, 2015).

Engineers Have Integrated Climate Change Adaptation into a Limited Number of Installation Projects Due to a Lack of Updated Design Standards

In the 2014 *Roadmap,* DOD stated that it will review—and as needed, modify—design and construction standards to account for the impacts of climate change. In addition, DOD guidance instructs components, including the military services, to design and construct proposed infrastructure projects using Unified Facilities Criteria standards. However, these standards are based on historical weather patterns and do not account for projected changes in the climate.[72] Thus, according to DOD officials and engineers responsible for the projects at the overseas locations that we visited or contacted, integration of climate change adaptation into the design of projects for both military construction and sustainment, restoration, and modernization has been limited because projects are typically designed to the Unified Facilities Criteria.[73]

During the course of our review, we identified some examples where installation engineers were able to integrate climate change adaptation into project designs, but this integration was generally inconsistent. For example, at a DOD installation in the Pacific, officials are developing plans for a vehicle maintenance facility that will be built in a floodplain. The officials explained that the engineers and planners plan to elevate the structure and incorporate water pumps to address the anticipated increase in flooding due to increased extreme weather events and storm surge. In another example, at a DOD installation in Europe, a gate separating the base from the community and providing anti-terrorism force protection had been damaged in several flooding events. As a result, a renovation project was designed to take into account previous flooding caused by increasingly severe rain events. The latest repair project modified the stream bed leading up to the gate in order to reduce the collection of storm water debris on the gate. For projects such as these, according to installation officials, engineers must receive approval on a case by case basis in order to modify project designs.

---

[72] According to DOD, the Unified Facilities Criteria program unifies all technical criteria and standards pertaining to planning, design, construction, and operation and maintenance of real property facilities.

[73] For all overseas DOD design and construction projects, military services must use the building standards found in the Unified Facilities Criteria, unless a more stringent Status of Forces Agreement, host-nation funded construction agreement, or bilateral infrastructure agreement applies. In this section of the report, we discuss infrastructure projects designed using the Unified Facilities Criteria. When collaborating with host nations on the design and construction of infrastructure, those projects use design standards found in both the Unified Facilities Criteria and host-nation design standards. We discuss those infrastructure projects in the next section of the report.

However, at most of the installations we visited or contacted, officials explained that engineers cannot design military construction and sustainment, restoration, and modernization projects beyond the Unified Facilities Criteria standards and thus are generally not integrating climate change adaptation into the design of installation projects. For example:

- According to DOD officials, climate change adaptation was not considered in the design of a $48.9 million project in Europe, expected to enhance the capabilities of an ammunition storage area located on a canal. The project is expected to enhance a dock, a bridge that will span the canal, and railroad tracks to transport ammunition. Officials noted that the canal is vulnerable to the risks of increased flooding due to sea level rise and high water events; but, there are no plans to modify the existing design to raise the dock or incorporate preventative measures to protect the railway. As a result, the operational capability of the storage area will be at risk to potential flooding, and may be in need of additional funding to address future modifications.

- According to officials at a DOD installation in the Pacific, sustainment, restoration, and modernization funds are used to repair infrastructure damaged by typhoons. The repairs include replacing doors around the base, but do not incorporate adaptation for the potential for increasingly strong winds.

- Furthermore, officials at a DOD installation in the Pacific noted that the Unified Facilities Criteria standards for storm water systems' drainage capacity do not account for projected increases in precipitation. Major rainstorm events in 2010 and 2014 caused large amounts of water to overwhelm the base's drainage system (see fig. 9) and mudslides. Also, flooding prevented guards from opening the front gates, preventing personnel from going to work. Officials stated that in order to integrate climate change adaptation into project designs, the Unified Facilities Criteria standards would need to include information that outlines specific climate change adaptation metrics to be integrated into installation-level designs and projects.

Figure 9: 2010 and 2014 Rainstorms Overwhelmed the Drainage System at a Department of Defense Installation in the Pacific

Source: Department of Defense. | GAO-18-206

DOD has taken some steps to produce certain climate change projection data that may influence these standards in the future or drive the need for additional guidance on the use of this emerging data. For example, a DOD weather unit creates projections of how certain climatic conditions at DOD overseas locations may change based on updated climate models.[74] In addition, in 2016, DOD made available to the military services a database with projections of future rising sea levels at over a thousand locations across the world, including at over two hundred overseas installations. However, military service and installation officials we visited or contacted stated that project designs are still governed by the Unified Facilities Criteria design standards; the use of current design standards continues to leave the military services vulnerable to the operational and budgetary risks associated with climate change impacts because they have not been updated. To provide installation officials with the flexibility they need to integrate climate change adaptation into projects, the standards would need to be updated and changed, according to these officials.

During the course of our review, we found that some military services are developing tools and guidance to instruct installation planners on

---

[74] According to unit documentation, the unit's current projections are based on climate change output from Coupled Model Intercomparison Project 5 models and, according to DOD officials, the unit plans to update the models using information from Coupled Model Intercomparison Project 6 models.

integrating climate change adaptation into infrastructure plans and projects as illustrated by the following examples.

- The Army is developing guidance on how to integrate climate change adaptation into Army installations' natural resources plans and is modifying a U.S. Army Corps of Engineers Civil Works vulnerability assessment tool that includes historic weather-related data and projections from authoritative sources.

- The Navy has taken three steps to provide additional instruction to installation-level planners on how to integrate climate change adaptation into planning efforts. First, the Navy is developing a guidebook and associated training to instruct planners on how to integrate climate change adaptation into natural resources plans. Second, the Navy developed the *Climate Change Planning Handbook: Installation Adaptation and Resilience* to help planners identify and assess adaption alternatives to manage potential impacts to current and planned infrastructure.[75] Third, the Navy has developed the *Sea Level Change Framework Report*, which summarizes the historical and emerging response to incorporating projected sea level change into infrastructure design projects.[76] In conjunction with this report, the Navy has developed a draft Engineering Construction Bulletin that provides guidance for incorporating information on sea level change into the design of structures to determine their elevation for flood protection in coastal areas.

However, according to our review of these efforts and military service officials, the efforts either are not planned for use by overseas installation planners or address only a limited number of the types of climate change impacts (mostly rising sea levels or flooding). As a result, the lack of design standards that reflect climate change hampers installation-level planners' and engineers' ability to consistently incorporate climate change adaptation into plans and individual projects. This potentially exposes the military services to increased operational and budgetary risks that are posed by weather effects associated with climate change.

---

[75]Leidos, Inc., and Louis Berger, Inc., *Climate Change Planning Handbook Installation Adaptation and Resilience,* a report prepared at the request of the Department of Defense, Naval Facilities Engineering Command Headquarters (Washington Navy Yard, D.C.: January 2017).

[76]AECOM, *Sea Level Change Framework Report,* a report prepared at the request of the Department of Defense, Naval Facilities Engineering Command (Sept. 30, 2016).

Climate Change Adaptation Is Generally Not Included in DOD's Collaboration with Host Nations at the National or Installation Level

DOD collaborates with host nations at both the national and installation level, but this collaboration and cost sharing generally does not include climate change adaptation. At the national level, DOD collaborates with some host-nation governments through three key mechanisms that implement cost-sharing activities, but these mechanisms generally do not include information on climate change adaptation. At the installation level, DOD collaborates with some host-nation communities in an effort to strengthen infrastructure resilience and share information on climate change, but the majority of installations we visited or contacted do not collaborate with host-nation communities on climate change adaptation. Without more fully including adaptation in its collaboration with host nations, DOD may miss opportunities to increase the resilience of host-nation-built infrastructure and installations to the operational and budgetary risks posed by weather effects associated with climate change.

DOD's Collaboration with Host Nations at the National Level Generally Does Not Include Climate Change Adaptation

According to both the 2010 and 2014 *Quadrennial Defense Reviews*, managing the national security effects of climate change will require DOD to work collaboratively with allies and further states that climate change impacts create both a need and an opportunity for nations to work together. Also, the 2016 *Sustainability Plan* stresses the importance of engaging with external stakeholders to, among other things, implement actions to maintain mission resilience in the face of a changing climate. In addition, DOD's 2014 *Roadmap* states that collaboration is essential to effectively adapt DOD plans and operations, and emphasizes cooperation with partner nations to enhance planning, responses, and resilience to the effects of climate change.

We found that DOD collaborates with some host-nation governments through three key types of collaborative mechanisms: national agreements, host-nation building standards, and negotiation processes. As discussed previously, these three collaborative mechanisms typically implement DOD's cost-sharing activities with host nations. However, these key mechanisms of national-level collaboration generally do not include information on ways to increase infrastructure resiliency through climate change adaptation.

- **Collaboration with host nations through national agreements:** Based on our review of State's 2016 Treaties in Force and interviews with officials from DOD's Office of the Assistant Secretary of Defense for Strategy, Plans, and Capabilities, DOD's Facilities Investment and Management Office, State's Offices of Congressional and Public Affairs and Security Negotiations and Agreements, and regional- and installation-level military service officials, there are no agreements between the United States and host nations in our sample relating to DOD cost-sharing activities that include information on climate change adaptation.[77]

- **Collaboration with host nations on building standards:** At the majority of installations we visited or contacted, military service officials told us that host-nation building standards do not include information on climate change adaptation.[78] As we discussed above, the building standards in DOD's Unified Facilities Criteria are similarly not based on climate change projections, but rather, on historical weather data. Since most host-nation-funded construction projects are designed and constructed using a combination of host-nation building standards and DOD's Unified Facilities Criteria, infrastructure projects funded through national cost-sharing activities are built without considering the operational and budgetary risks posed by weather effects associated with climate change.[79] For example, we visited two DOD locations in the Pacific with planned host-nation-funded construction projects to repair seawalls protecting ammunition depots (see fig. 10). However, according to DOD officials, these repair projects do not account for a potential increase in average sea levels

---

[77]We reviewed State's 2016 Treaties in Force and found two agreements that referred to climate change, including a 2010 Memorandum of Understanding between the U.S. Department of the Interior and the Socialist Republic of Vietnam concerning Scientific and Technical Cooperation in Earth Sciences and Effective Management of Natural Resources in the Context of Climate Change and a 1994 United Nations Framework Convention on Climate Change. These two agreements did not specifically address climate change adaptation for infrastructure used by DOD overseas. The Treaties in Force publication is prepared by State for the purpose of providing information on treaties and other international agreements to which the United States has become a party and that are carried on State's records as being in force as of its stated publication date, Jan. 1, 2016.

[78]While the majority of host-nation building standards for the countries in our sample do not include information on climate change adaptation, several military service officials told us that the building standards for the United Kingdom and Germany incorporated climate change adaptation and allowances necessary for the local area, as required by host-nation regulations. Officials at these installations collaborated with host-nation officials to ensure infrastructure projects adhered to the applicable host-nation building standards.

[79]While most host-nation building standards for the countries in our sample do not include information on climate change adaptation, we learned about a few planned or completed host-nation-funded construction projects that incorporated adaptation features. For example, a DOD installation in the Pacific is collaborating with host-nation officials on several projects to raise the height of the installation's seawall. According to officials, while these host-nation-funded projects do not incorporate projections of future rising sea levels for that location, the increased height of the seawall will still protect against higher storm surge if typhoons become more severe in the future.

because this information is not included in host-nation building standards or the Unified Facilities Criteria. Given the potential for future rising sea levels and an associated storm surge, access to the depots may be limited during training and contingency operations, according to installation officials.

Figure 10: Damaged Seawall at a Department of Defense Ammunition Depot in the Pacific

Source: GAO. | GAO-18-206

- **Collaboration with host nations through negotiations:** The negotiation process and preparations for negotiations between the United States and host nations on national agreements and cost-sharing activities have not included information on climate change adaptation, according to DOD and State officials. Instead, according to officials from the sub-unified commands, DOD's Office of the Assistant Secretary of Defense for Strategy, Plans, and Capabilities, and State, negotiations and preparations for negotiations between the United States and host nations in our sample mainly focus on funding contributions for national cost-sharing activities. Negotiations can also include discussions on the process for designing projects, such as the use of host-nation building standards in projects funded via cost-sharing activities. DOD and State officials told us that climate change adaptation has not been included in preparations for negotiations; consequently, adaptation efforts have not been incorporated into national cost-sharing activities or agreements with host nations, including the host-nation building standards used in cost-sharing activities.

As noted above, both the 2010 and 2014 *Quadrennial Defense Reviews* state that managing the national security effects of climate change will require DOD to work collaboratively with allies and that climate change impacts create both a need and an opportunity for nations to work together. In addition, according to the 2016 *Sustainability Plan*, DOD plans to engage with external stakeholders to implement actions that maintain mission resilience in the face of a changing climate. Also, DOD's 2014 *Roadmap* states that collaboration is essential to effectively adapt DOD plans and operations, and emphasizes cooperation with partner nations to enhance planning, responses, and resilience to the effects of climate change. Further, as discussed above, installation officials explained that costs often associated with climate change impacts are for the sustainment, restoration, or maintenance of installation infrastructure. Since, according to installation officials, DOD assumes responsibility for sustainment and maintenance costs once host nations have completed construction for infrastructure funded through national cost-sharing activities, DOD is vulnerable to budgetary risks associated with climate change. Integrating climate change adaptation into the three key collaborative mechanisms of cost-sharing—national agreements, host-nation building standards, and negotiation processes—presents an opportunity for DOD to more comprehensively address climate change. More specifically, some DOD officials told us that information on climate change adaptation could be included in the negotiation process, including preparations for negotiations, to ensure cost-sharing activities consider the operational and budgetary risks associated with climate change. For example, officials at a DOD installation in the West Indies told us they are interested in negotiating an expanded agreement with the host nation where information sharing on climate change and resiliency could be explored. Without considering climate change adaptation as DOD prepares to negotiate on cost-sharing activities and agreements, DOD will continue to use host-nation-built infrastructure that may be vulnerable to the operational and budgetary risks posed by weather effects associated with climate change and may miss opportunities to increase infrastructure resilience to these risks.

DOD's Collaboration with Host Nations at the Installation Level Generally Does Not Include Climate Change Adaptation

DOD's installation-level collaboration with host-nation communities generally does not include discussions of climate change adaptation. These partnerships, according to DOD's 2014 *Roadmap*, are needed to fully ensure the department's mission is sustainable under a changing climate. In addition, the 2014 *Roadmap* states that the department's decisions—and those of neighboring communities—are intrinsically interconnected; and that the department cannot effectively assess its vulnerabilities and implement adaptive responses at its installations if neighbors and stakeholders are not part of the process. Further, the 2014 *Roadmap* states that DOD will enhance collaboration via cooperation with partner nations, host-nation military, and other appropriate organizations on planning, responses, and resilience to the effects of climate change. Moreover, the 2014 *Roadmap* states that effective collaboration with internal and external stakeholders includes collaboration with surrounding communities for planning climate change adaptation and emergency preparedness and response. Also, in the 2016 *Sustainability Plan*, DOD stresses the importance of the military services' installations collaborating with surrounding communities to adapt to the risks posed by climate change.

However, based on our review, DOD officials at 32 out of 45 installations that we visited or contacted told us they do not collaborate on climate change adaptation with host-nation communities or officials. The remaining thirteen installations had engaged in collaborative efforts, which included strengthening the resiliency of infrastructure used by DOD, used by both DOD and the host nation, or used solely by the host nation, as well as improving information sharing on climate change.[80] For example, we found:

- **Collaboration can strengthen the resiliency of infrastructure used by DOD:** Two DOD installations in Europe collaborated with surrounding communities on flooding caused by heavy rain events that impacted DOD infrastructure. One installation collaborated with local city officials on a project to modify a creek bed in the neighboring community. During heavy rain events, large amounts of sediment, gravel, and debris traveled down this creek, damaging the installation's anti-terrorism force protection gate and causing flooding on base. DOD officials, working with local host-nation officials,

---

[80]For the purposes of this report, strengthening infrastructure "resilience" refers to making infrastructure more resilient to the operational and budgetary risks posed by weather effects associated with climate change.

widened and paved the creek bed outside the installation's fence line, which reduced flooding on base and protected the gate from further damage. Officials at a DOD installation in the area assigned to the U.S. Africa Command also told us that they have engaged in discussions with local government officials on ways to increase the resiliency of local port infrastructure through which all major U.S. cargo arrives at the installation. According to officials, the pier and associated shipping warehouses are at risk of future rising sea levels and in need of a redesign to support safe and modern shipping. Installation officials stated that they regularly communicate and collaborate with host nation government personnel on construction projects, such as pier redesign, and other mutual support activities to assure mission success. As a result, DOD infrastructure will be more resilient to the operational and budgetary risks posed by weather effects associated with climate change in the future.

- **Collaboration can strengthen the resiliency of infrastructure used by both DOD and the host nation:** Regular flooding at a petroleum, oil, and lubricant facility in the Pacific limits DOD access and blocks traffic for the local community. During severe rain storms, debris clogs the drainage tunnels and gates and causes storm water to flood both the facility and the local highway. In these cases, DOD officials said they cannot access and pump jet fuel to an adjacent DOD installation until flooding has receded. DOD officials told us this flooding is a severe impact to the local community because the road becomes impassable during heavy rain events. According to DOD officials, they are collaborating with local host-nation officials on two projects to repair and replace DOD infrastructure at the facility in 2017, enhancing the infrastructure used by both DOD and local community members. Also, a DOD installation in the Pacific began collaborating with city and regional government officials in 2016 on flooding issues directly caused by DOD installation infrastructure. During heavy rain events, flooding on the installation's runway shuts down flight operations, limits access to parts of the base (e.g., the hospital), and causes flooding in the surrounding host-nation community. Installation officials are currently working with the regional host-nation government and the Mayor of the surrounding community to request a host-nation funded construction project to build a detention pond inside the installation boundary. This project would not only increase the resiliency of infrastructure used by DOD, but also the infrastructure of the surrounding community.

- **Collaboration can strengthen the resiliency of infrastructure used by the host nation:** A DOD installation in the Pacific completed two erosion mitigation projects in 2013 and 2015 to address host-nation community concerns about reoccurring landslides limiting access on local roads. Installation officials told us these erosion problems were on hillsides outside the installation fence line, but still technically located on DOD property. After meeting with local community members, installation officials used DOD funds to stabilize hillside erosion. According to officials, these two projects helped to maintain good relationships with local city officials.

In addition, collaboration between installation and host-nation officials can also lead to climate change information sharing. For example, a DOD installation in Europe participates in a climate change working group with city and regional government officials. This working group aims to support members in evaluating the potential impacts of climate change to their mission; establishing networks between stakeholders to share information and develop local mitigation strategies; and cooperating with stakeholders to implement mitigation strategies.

As the examples above indicate, opportunities exist for DOD to expand collaborative efforts with host-nation communities at overseas installations in order to increase infrastructure resilience to the operational and budgetary risks posed by weather effects associated with climate change. However, according to DOD officials, installation-level collaboration with surrounding communities generally does not include climate change adaptation, in part due to a lack of formal coordination requirements and implementing guidance. In our prior work on best practices for collaboration, we found that formal written guidance and agreements, such as memorandums of understanding (memorandums), strengthen collaboration.[81] A majority of the installations in our sample that have collaborated with surrounding host-nation communities on mutual aid response have done so through memorandums.[82] These memorandums have been a tool for enhancing installations' resilience to emergencies, including those associated with extreme weather events,

---

[81]GAO-12-1022.

[82]DOD defines mutual aid as reciprocal assistance by emergency services under a prearranged agreement or plan. DODI 6055.06, *DOD Fire and Emergency Services (F&ES) Program*, para. E2.32 (Dec. 21, 2006). For example, with an established agreement, DOD could respond to and provide services to address a fire in the surrounding community. Similarly, the surrounding community could respond to a fire located on the actual DOD installation. The memorandums provided by the installations in our sample covered a wide range of support activities, including fire response and protection, disaster preparedness and response, general aid and ambulatory services, and emergency management cooperation.

but this tool is not being used to respond to climate change adaptation. Installation officials in Europe and the Pacific told us that memorandums are an appropriate and effective type of agreement for enhancing an installation's resilience to the impacts of extreme weather and that the establishment of a memorandum does not require considerable effort on DOD's part, but that currently, these are not being used as a tool to respond to climate change adaptation. The Chief of Emergency Management at a DOD installation in Europe explained that installation and host-nation officials are more likely to engage in collaboration if that specific collaboration is described and required through a formal arrangement such as a memorandum. Without consistent collaboration on climate change adaptation with host-nation communities and officials, the services may miss opportunities to increase installations' resilience to the operational and budgetary risks posed by weather effects associated with climate change. Taking steps to ensure that DOD officials have formal guidance on collaborating with host-nation communities could help installations take advantage of opportunities to strengthen the resilience of infrastructure used by DOD overseas.

Conclusions

DOD policy states that the U.S. foreign and overseas posture is the fundamental enabler of U.S. defense activities and military operations overseas and is also central to defining and communicating U.S. strategic interests to allies, partners, and adversaries.[83] Further, the department's overseas infrastructure provides facilities and training and testing ranges that are vital to the department's ability to fulfill its mission and constitutes a significant fiscal commitment that requires ongoing investment to maintain. In its 2010 and 2014 *Quadrennial Defense Reviews* and 2016 *Sustainability Plan*, DOD states that climate change will have serious implications for the ability of the department to maintain its infrastructure and ensure military readiness.

DOD installations overseas have experienced operational and budgetary risks posed by weather effects associated with climate change impacts at the military services' installations in each of DOD's geographic combatant commands. The observed effects associated with climate change have already negatively impacted the installations' ability to execute key training and testing operations and missions. Further, based on projections of future climate change, these negative impacts are likely to

---

[83]DOD Instruction 3000.12.

increase, and DOD has recognized the need to adjust its modeling of future maintenance and sustainment costs for infrastructure to account for the budgetary risks posed by climate change. To do so, however, DOD needs data on the costs associated with the impacts of climate change so that the department can modify its modeling in an accurate way. While the military services and their installations are generally able to track certain types of costs, the installations do not consistently track these costs because the military services do not require it. As a result, DOD lacks information needed to adjust its budgeting for the increased maintenance and sustainment costs that are likely to come with climate change.

To manage the operational and budgetary risks posed by weather effects associated with climate change, DOD surveyed its overseas installations on climate change impacts, such as flooding and changes in sea level, but the information DOD collected was not complete or comprehensive. Without complete and comprehensive data on the risks posed by weather effects associated with climate change, the military services will not have the information they need to support DOD's plans to incorporate identified risks into planning and risk management processes.

In addition, DOD modified its guidance for master plans and natural resources plans, adding more detail to how these plans should address climate change adaptation. Although these requirements have been in place since 2012 and 2011, respectively, most plans we reviewed did not address climate change adaptation. Lack of (1) consistent communication on the requirements to integrate climate change adaptation into installation plans, (2) instruction to planners to do so through service-level guidance, and (3) training on the development of installation-level plans has contributed to this issue. Moreover, the department has not updated its Unified Facilities Criteria design standards to require the integration of climate change projections. Updated design standards that account for a changing climate could help provide engineers with the information they need to design infrastructure that is more resilient to the operational and budgetary risks posed by weather effects associated with climate change.

Finally, although host nations bear the initial cost of overseas infrastructure at some locations, DOD bears the cost of sustainment and maintenance once host nations complete construction, which can be for a period of 50 or more years. However, DOD has not included climate change adaptation in negotiations with host nations or, generally, in partnerships with communities and is missing an opportunity to assist U.S. allies in making this infrastructure more resilient to the operational and budgetary risks posed by weather effects associated with climate change. The lack of formal memorandums, that could provide guidance to promote enhancement of resilience to the operational and budgetary risks posed by weather effects associated with climate change, also impacts these collaborative efforts.

Recommendations for Executive Action

We recommend that the Secretaries of the Army, Navy, and Air Force take the following four actions:

- Work with the Office of the Secretary of Defense to issue a requirement to their installations to systematically track the costs associated with extreme weather events and climate change. (Recommendation 1)

- Take steps to administer the Screening Level Vulnerability Assessment Survey, or a similar instrument, to all relevant locations. (Recommendation 2)

- Implement DOD goals and plans by incorporating climate change adaptation into service-level guidance and required training for the development of installation-level plans, including master plans and natural resource plans, at all locations. (Recommendation 3)

- Integrate climate change data and projections into the Unified Facilities Criteria standards and periodically revise those standards based on any new projections, as appropriate. (Recommendation 4)

We also recommend that the Secretary of Defense take the following two actions:

- Direct the Office of the Under Secretary of Defense (Policy), the geographic combatant commands, the sub-unified commands, and the Secretaries of the Army, Navy, and Air Force to consider climate change adaptation as they develop DOD's position for future negotiations with host-nation governments on cost-sharing activities, when relevant or appropriate. (Recommendation 5)

- Direct the Secretaries of the Army, Navy, and Air Force to issue guidance, as appropriate, that calls for more formal coordination mechanisms related to climate change adaptation, such as memorandums of understanding, between DOD installations and surrounding host-nation communities. (Recommendation 6)

Agency Comments and Our Evaluation

We provided a draft of this report for review and comment to DOD and State. In written comments, DOD partially concurred with four recommendations and non-concurred with two. DOD's comments are summarized below and reprinted in their entirety in appendix II. In an e-mail, State indicated that it did not have formal comments. In addition, DOD and State provided technical comments, which we incorporated as appropriate. Based on technical comments received from DOD, we revised the addressees of several recommendations. Specifically, for the recommendations where we directed the action to the Secretaries of the Army and Air Force, the Chief of Naval Operations, and the Commandant of the Marine Corps, we are now directing the actions to the Secretaries of the Army, Navy, and Air Force because the Navy noted that the Secretary of the Navy fully represents the Chief of Naval Operations and the Commandant of the Marine Corps.

In its overall comments, DOD stated that the report cites a DOD position or policy that, in numerous cases, is neither current nor accurate, and that the report repeatedly cites DOD documents and threat reviews as though they are current as of July 2017. DOD also stated that using the term "According to DOD" without a specific cite, attribution, or context is misleading and should not be included in a professional report. Further, DOD stated in its comments that ascribing infrastructure damage specifically to climate change impacts without taking into account the effects of extreme weather events is speculative at best and misleading, noting that GAO has found in previous reports that it is not possible to link any individual weather event to climate change. Finally, DOD stated that the department is in the process of updating a National Military Strategy and National Defense Strategy to focus resources on threats considered to be critical to our Nation's security and the security of our Allies.

We disagree with several of DOD's statements. First, with regard to DOD positions or policies cited in this report, according to senior officials in the office of the Assistant Secretary of Defense for Energy, Installations, and the Environment, DOD is in the process of updating guidance and making revisions to its policy on climate change. We recognize DOD's ongoing efforts to review its climate-related guidance and policies. However, the department had not made any official revisions to its policy as of October 27, 2017. Moreover, as our report states, DOD has emphasized the importance of climate change in key strategy, policy, and guidance documents and reports since 2010. Specifically, our report states that DOD has emphasized the importance of the threat from climate change in its 2010 and 2014 Quadrennial Defense Reviews, the current National Military Strategy, the 2012 Strategic Sustainability Performance Plan (and each subsequent annual version of this plan), and its 2014 Arctic Strategy. According to an official from the Office of the Assistant Secretary of Defense for Energy, Installations, and the Environment, none of these documents have been withdrawn by DOD nor otherwise repudiated as of October 27, 2017.

Further, DOD has issued additional guidance and policy documents that, as our report states, focus on the threat posed to national security by climate change. These documents include the 2014 Climate Change Adaptation Roadmap and DOD Directive 4715.21, which remain unrevised as of October 27, 2017. Senior DOD officials have also cited the department's position on climate change in congressional testimony. For example, as our report states, in March 2016 testimony before the House Appropriations Committee, then acting Assistant Secretary of Defense for Energy, Installations and Environment stated that DOD would continue its efforts to develop the science and tools needed to meet the department's obligations to assess and adapt to climate change. The Assistant Secretary (Acting) reiterated that resilience to climate change continued to be a priority for DOD, explaining that—even without knowing precisely how or when the climate will change—the department knows it must build resilience into its policies, programs, and operations in a thoughtful and cost effective way.[84]

With regard to DOD's assertion that using the term "according to DOD" without a specific cite, attribution, or context is misleading and should not be included in a professional report, our use of the term "according to DOD" is accurate throughout the report and reflects information provided to us by DOD officials acting in their official capacities, and we provide specific attribution, as appropriate, to the source of the information we are discussing. Examples of sources include installations from which we collected information, which are known to the Office of the Assistant Secretary of Defense for Energy, Installations, and the Environment, as well as DOD policy, guidance, and reports that address climate change.

In addition, we disagree that our report ascribes infrastructure damage specifically to climate change impacts without taking into account the effects of extreme weather events. As we stated in a 2014 report, and

---

[84]Installations, Environment, Energy and BRAC: Hearing Before the Subcomm. on Military Construction, Veterans Affairs and Related Agencies of the H. Comm. on Appropriations, 114th Cong. (2016) (statement of Pete Potochney, performing the duties of Assistant Secretary of Defense (Energy, Installations and Environment)).

again in this report, it is not possible to link any individual weather event to climate change. In this report, we use the phrase "weather effects associated with climate change" to characterize the relationship between climate change and individual weather events. Our characterization parallels that of DOD. As our report states, DOD's 2016 draft report on the department's Screening Level Vulnerability Assessment uses the term "seven effects commonly associated with climate change"—which are summarized in Table 1— while DOD's Roadmap uses a similar phrase— "climate-related effects." Our characterization is also consistent with findings from the National Academy of Sciences and other government or scientific organizations.

Consistent with our previous work, we note in this report that while it is not possible to link any individual weather event to climate change, these events provide insight into the potential climate-related vulnerabilities that DOD has reported it faces. In this report, we reference DOD policy documents, plans, and reports that provide specific examples of events that the department has associated with climate change. These documents include the department's Quadrennial Defense Review, Strategic Sustainability Performance Plan, and the Roadmap. We also note that DOD installation-level officials with whom we met identified vulnerabilities in the department's facilities and infrastructure associated with climate change phenomena. Specifically, as our report states, these installation-level officials recognized that climate change may make these types of phenomena more frequent or severe and that impacts of weather effects associated with climate change pose operational risks to training, testing, and mission operations, as well as both incremental and one-time costs.

DOD non-concurred with our recommendation that the Secretaries of the Army, Navy, and Air Force work with the Office of the Secretary of Defense to issue a requirement to their installations to systematically track the costs associated with extreme weather events and climate change. In its response, DOD stated that tracking impacts and costs associated with extreme weather events is important, but that the science of attributing these events to a changing climate is not supported by previous GAO reports. Also, DOD noted that currently, associating a single event to climate change is difficult and does not warrant the time and money expended in doing so.

We continue to believe that our recommendation is appropriate. DOD officials from the military services' overseas regional organizations and installations in our sample explained that installations generally have the capability to track these types of costs associated with extreme weather events that are projected to become more frequent and intense due to climate change. As we state in this report, there is—according to DOD installation officials—substantial budgetary risk resulting from weather effects associated with climate change, spread across each of the combatant commands. Moreover, we have previously reported that these types of repairs are neither budgeted for nor clearly represented in the federal budget process.[85] Also, DOD's Roadmap emphasizes the resource implications of climate change. For example, the department noted that changing building heating and cooling demand could impact installation energy intensity and operating costs and that facility maintenance and repair cost models may need to be modified accordingly. Moreover, our report documents the substantial costs of weather effects associated with climate change at several overseas installations. Finally, as we state in this report, if the military services do not track these costs, they will lack the information they need to adapt infrastructure at overseas installations to weather effects associated with climate change and be unlikely to develop accurate budget estimates for infrastructure sustainment.

DOD non-concurred with our recommendation that the Secretaries of the Army, Navy, and Air Force take steps to administer the Screening Level Vulnerability Assessment Survey (SLVAS), or a similar instrument, to all relevant locations. In its response, the department stated that SLVAS is an internally-developed survey instrument intended to provide an initial broad-level screening of climate impacts across DOD. The department further noted that SLVAS does not provide quantitative data or account for mission criticality, and therefore is not a useful tool for long-term decision making. DOD stated that it will encourage the military departments to administer the SLVAS—or an instrument they deem appropriate—as appropriate and as resources permit.

We recognize that SLVAS resulted in qualitative, not quantitative, analysis of climate-related vulnerabilities. Nonetheless, DOD developed the survey to identify DOD installations worldwide with climate-related vulnerabilities. As DOD's draft report notes, the survey's qualitative questions were framed to begin to identify sites with current weather-related effects and where more comprehensive assessment may be needed in order to identify potential effects of future climate change. DOD

---

[85]GAO-17-317.

reported that SLVAS responses yielded a wide range of qualitative information, including a preliminary picture of assets currently affected by severe weather events, as well as an indication of assets that may be affected by sea level rise in the future. However, DOD does not explain in its response why qualitative information is not useful. DOD recognized that SLVAS was a first step in an on-going process to manage the risks to DOD's mission associated with a changing climate. Further, DOD previously stated that the survey could serve as a useful tool to inform long-term decision making. Moreover, as our report notes, a significant portion of DOD's most strategically important overseas sites were not surveyed. Surveying these locations would allow DOD to more accurately account for climate-related vulnerabilities at a wider range of locations, including all of those with critical missions. During the course of our review, DOD officials stated that the department had acknowledged these challenges and was planning to determine the appropriate course of action to capture pertinent information in the future. If DOD believes it needs to make changes in the SLVAS survey or its implementation, or wishes to develop another means of obtaining relevant data, that would likely meet the intent of our recommendation.

DOD partially concurred with our recommendation that the Secretaries of the Army, Navy, and Air Force should implement DOD goals and plans by incorporating climate change adaptation into service-level guidance and required training for the development of installation-level plans, including master plans and natural resource plans. DOD stated that it plans to revise Directive 4715.21, which will provide goals and requirements for all DOD components. However, DOD did not address providing training for the development of installation-level plans, including master plans and natural resource plans in its response. If DOD provides both goals and requirements for all DOD components and training for the development of relevant plans, the department's response will likely meet the intent of the recommendation.

DOD partially concurred with our recommendation that the Secretaries of the Army, Navy, and Air Force should integrate climate change data and projections into the Unified Facilities Criteria standards and periodically revise those standards based on any new projections, as appropriate. DOD stated that the Office of the Secretary of Defense plans to work with the military departments to evaluate the Unified Facilities Criteria to determine where and when the use of climate data is appropriate, and modify and publish Unified Facilities Criteria if warranted. As our report states, DOD has already taken steps to produce certain climate change projection data, including a DOD weather unit that creates projections of certain climatic conditions at overseas locations and a database with projections of future rising sea levels at more than 1,000 locations across the world, including at more than 200 overseas installations. If effectively implemented, to include incorporation of any current and relevant projections as well as new projections, DOD's plans would likely meet the intent of our recommendation.

DOD partially concurred with our recommendation to consider climate change adaptation as it develops its position for future negotiations with host-nation governments on cost-sharing activities, when relevant or appropriate. DOD stated that it will work to review the processes and criteria governing host-nation cost-sharing negotiations to strengthen or incorporate consideration of resilience when relevant or appropriate. DOD also stated that each bilateral agreement is different and individually negotiated and thus, a universal solution is not expected. We recognize that each bilateral agreement is individually negotiated and unique to the specific host nation. If effectively implemented, DOD's actions would likely be responsive to our recommendation.

DOD partially concurred with our recommendation to issue guidance, as appropriate, that calls for more formal coordination mechanisms related to climate change adaptation between DOD installations and surrounding host-nation communities. DOD stated that it will work to review the guidance for establishing agreements between host-nation communities and DOD installations. The department also stated that if necessary, it will identify where it would be appropriate to incorporate consideration of resilience. If DOD implements needed changes to provide for formal coordination mechanisms or some other process that ensures coordination, and the department's consideration of resilience also includes consideration of climate change adaptation, these actions would likely meet the intent of our recommendation.

As agreed with your offices, unless you publicly announce the contents of this report earlier, we plan no further distribution until 30 days from the report date. At that time, we will send copies to appropriate congressional committees; the Secretary of Defense; the Secretaries of the Air Force, Army, and Navy; the Commandant of the Marine Corps; and the Secretary of State. In addition, the report will be available at no charge on the GAO website at http://www.gao.gov.

If you or your staff have any questions about this report, please contact me at (202) 512-4523 or leporeb@gao.gov. Contact points for our Offices of Congressional Relations and Public Affairs may be found on the last page of this report. GAO staff who made major contributions to this report are listed in appendix III.

Brian J. Lepore
Director, Defense Capabilities and Management

List of Requesters

The Honorable Jack Reed
Ranking Member
Committee on Armed Services
United States Senate

The Honorable Bernard Sanders
Ranking Member
Committee on the Budget
United States Senate

The Honorable Ben Cardin
Ranking Member
Committee on Foreign Relations
United States Senate

The Honorable Tim Kaine
Ranking Member
Subcommittee on Readiness and Management Support
Committee on Armed Services
United States Senate

The Honorable Richard Durbin
Ranking Member
Subcommittee on Defense
Committee on Appropriations
United States Senate

The Honorable Brian Schatz
Ranking Member
Subcommittee on Military Construction, Veterans Affairs, and Related Agencies
Committee on Appropriations
United States Senate

Appendix I: Scope and Methodology

To examine the extent to which the Department of Defense (DOD) has identified the operational and budgetary risks posed by weather effects associated with climate change on infrastructure used by DOD overseas, we first reviewed data collected by DOD's Screening Level Vulnerability Assessment Survey (survey) for 2013 through 2015. These data allowed the Office of the Secretary of Defense and the military services to work together to assess certain climate-related vulnerabilities at both domestic and overseas installations and associated sites. Using these data, we developed a nongeneralizable sample of 45 overseas installations. We interviewed military service officials and collected documentation on the observed impacts of extreme weather events and climate change, as well as associated costs, at these installations. To observe both the physical impacts of extreme weather events and climate change on infrastructure and adaptation or resilience measures taken or planned at the installation level,[1] we visited 14 installations in our sample.[2] Further, we reviewed DOD's 2010 and 2014 *Quadrennial Defense Reviews*, DOD's 2014 *Climate Change Adaptation Roadmap* (2014 *Roadmap*), DOD's Reports to Congress on Sustainable Ranges from 2011 to 2016, and DOD's *Strategic Sustainability Performance Plans* from 2012 to 2016 to identify climate change phenomena and effects.[3] We then interviewed installation officials about the types of costs associated with the risks posed by weather effects associated with climate change, such as facility maintenance and repair costs, and how often DOD installations track these costs. We compared what installation officials told us about tracking these costs with DOD Directive 4715.21, which requires DOD components to incorporate climate change resource considerations related to adapting built and natural infrastructure to potential or observed

---

[1]Adaptation includes considerations of climate change, such as whether or not specific adaptation actions are necessary, based on risk. Climate change adaptation differs from mitigation, which is focused on reducing emissions. In this report, we focus on DOD adaptation efforts.

[2]We visited 14 of the 45 foreign locations in our sample. We did not visit or contact any installations located within the United States and its territories.

[3]DOD, *Quadrennial Defense Review Report* (Washington, D.C.: Feb. 1, 2010); *Quadrennial Defense Review 2014* (Washington, D.C.: Mar. 4, 2014); *Strategic Sustainability Performance Plan FY 2016* (Sept. 7, 2016); *2014 Climate Change Adaptation Roadmap* (Alexandria, VA: June 2014) (hereinafter cited as DOD, 2014 *Roadmap*). See, e.g., Secretary of Defense, Under Secretary of Defense for Personnel and Readiness, *2015 Report to Congress on Sustainable Ranges* (March 2015).

Appendix I: Scope and Methodology

climate change impacts into installation-level planning efforts.[4] Also, we compared the information from these officials with the *Standards for Internal Control in the Federal Government* and Executive Order 13693, which state—respectively—that an agency's managers should use quality information to achieve the agency's objectives and that the head of each agency shall ensure that agency operations and facilities prepare for impacts of climate change by—among other actions—calculating the potential cost and risk to mission associated with agency operations.[5]

To examine the extent to which DOD has collected data to effectively manage the operational and budgetary risks of weather effects associated with climate change to overseas infrastructure, we reviewed DOD guidance, to include DOD Directive 4715.21, to understand the military services' roles and responsibilities for assessing climate change impacts on infrastructure. We also reviewed DOD's 2010 and 2014 *Quadrennial Defense Reviews* as well as DOD's 2012 and 2014 *Climate Change Adaptation Roadmaps* to determine DOD's goals for conducting vulnerability assessments. In addition, we reviewed DOD's 2016 *Strategic Sustainability Performance Plan* (2016 *Sustainability Plan*), which addresses the department's approach to the management of the risks posed by climate change. Further, we reviewed our previous work on how governments can best use information to manage the risks posed by climate change and our previous work on integrating climate change adaptation into civilian building codes, to gain insight into the time required for such integration.[6] We also reviewed guidance from the Office of the Secretary of Defense that accompanied the administration of DOD's survey, which required the military services to survey their installations about the risks posed by weather effects associated with

---

[4]DOD Directive 4715.21, *Climate Change Adaptation and Resilience* (Jan. 14, 2016). DOD Directive 4715.21 was issued in accordance with the direction in Executive Order 13653. On March 28, 2017, the Presidential Executive Order on Promoting Energy Independence and Economic Growth rescinded Executive Order 13653. According to an official from the Office of the Secretary of Defense, as of May 2017, DOD was working to determine the course of action the department will take with regard to its directive.

[5]GAO, *Standards for Internal Control in the Federal Government*, GAO-14-704G (Washington, D.C.: Sept. 10, 2014); Executive Order No. 13693, *Planning for Federal Sustainability in the Next Decade*, 80 Fed. Reg. 15869 (Mar. 19, 2015).

[6]GAO, *Improved Federal Coordination Could Facilitate Use of Forward-Looking Climate Information in Design Standards, Building Codes, and Certifications*, GAO-17-3 (Washington, D.C.; Nov. 30, 2016) and *Climate Information: A National System Could Help Federal, State, Local, and Private Sector Decision Makers Use Climate Information*, GAO-16-37 (Washington, D.C.: Nov. 23, 2015).

Appendix I: Scope and Methodology

climate change, and survey instructions that the department provided to the military services, along with best practices for conducting surveys.[7] Among other things, this guidance was developed to aid survey completion and included information on the exemption and distribution of the survey to installations and sites. To gather information on the impacts observed by DOD personnel, and associated costs, we interviewed and reviewed documentation from DOD officials in the Office of the Under Secretary of Defense for Acquisition, Technology, and Logistics; the Office of the Under Secretary of Defense for Intelligence; the Joint Staff; the headquarters of the Army, Navy, Marine Corps, and Air Force; the geographic combatant commands and their regional service components; the sub-unified commands; and the military services' installations in our sample. Finally, we reviewed DOD's 2016 Enduring Locations Master List to identify foreign infrastructure of particular significance to DOD missions.

To examine the extent to which DOD integrated adaptation to weather effects associated with climate change into its installation planning and project design efforts, we reviewed guidance requiring or stressing the need for DOD components to integrate climate change adaptation into certain installation and infrastructure planning efforts, including DOD Directive 4715.21, DOD Instruction 4715.03 on the department's Natural Resources Conservation Program, DOD's 2014 *Roadmap*, and the 2012

---

[7]Through the survey, DOD installations could report the following climate change impacts: drought, extreme temperatures (hot or cold), flooding and other impacts due to non-storm surge events, flooding due to storm surge, implications of increased mean sea level, wildfire, and wind. For a discussion of best practices for conducting surveys, see GAO, *The Democratic Republic of the Congo: Information on the Rate of Sexual Violence in War-Torn Eastern DRC and Adjoining* Countries, GAO-11-702 (Washington, D.C.: July 2011).

Appendix I: Scope and Methodology

Unified Facilities Criteria (2-100-01) on Installation Master Planning.[8] We also reviewed plans (i.e., installation master plans, natural resources management plans, and encroachment management plans) from the installations in our sample and assessed these plans against DOD policy for incorporating climate change adaptation into installation planning efforts. This included the department's 2014 memorandum on installation floodplain management, which states that the DOD must plan and manage those facilities vulnerable to climate-related flooding and DOD's 2015 guidance on sustaining access to training areas, which instructs planners to evaluate the risks to training and range capability from the impacts of climate change trends.[9] We reviewed project documentation for proposed or approved installation-level military construction projects in our sample to determine the extent to which DOD is integrating climate change adaptation into its foreign infrastructure and compared this information with DOD guidance. Similarly, we discussed adaptation efforts funded with operations and maintenance or sustainment, restoration, and modernization funds with DOD officials to determine whether these efforts incorporated climate change adaptation. We interviewed DOD officials at the military services' headquarters and at the selected installations to determine the extent to which the services have implemented climate change adaptation efforts at the installation level. We also interviewed DOD officials from the Office of the Under Secretary of Defense for Acquisition, Technology, and Logistics, the geographic combatant commands and their regional service components, and the

---

[8]DOD, Unified Facilities Criteria 2-100-01, *Installation Master Planning* (May 15, 2012) states that installation planners can prepare a master plan that sustainably accommodates future change by incorporating current needs and mission requirements into a vision with clear goals and measurable objectives. The guidance further states that the military services' master planners are to understand, monitor, and adapt to, among other things, changing climatic conditions. We also reviewed DOD Instruction 4715.03 governing the department's Natural Resources Conservation Program on domestic installations. According to DOD officials, the military departments have chosen to use the instruction as guidance for their overseas installations' development of these plans and installations in our sample have used the instruction in this way. The instruction states that all DOD natural resources conservation programs shall be integrated with installation planning and programming. The guidance further states that, for natural resources plans, all DOD components are to utilize existing tools to assess the potential impacts of climate change to natural resources on DOD installations, to the extent practicable and using the best science available. DOD, Instruction 4715.03, *Natural Resources Conservation Program* (Mar. 18, 2011).

[9]Office of the Under Secretary of Defense for Acquisition, Technology, and Logistics Memorandum, *Floodplain Management on Department of Defense Installations* (Feb. 11, 2014); DOD Instruction 3200.21, *Sustaining Access to the Live Training Domain* (Sept. 15, 2015).

Appendix I: Scope and Methodology

sub-unified commands about installation-level planning efforts and planned or completed climate change adaptation projects.

To examine the extent to which DOD has collaborated with host nations on adapting infrastructure used by DOD to increase resiliency to the impacts of weather effects associated with climate change and shared costs for any needed adaptation, we collected information from DOD and Department of State officials on collaboration between DOD and host nations on climate change adaptation and cost-sharing activities. We compared this information with DOD Directive 4715.21 and DOD's 2014 Roadmap, to learn about DOD's requirements and goals for collaboration with external stakeholders to address climate change challenges. We reviewed bilateral agreements between DOD and host-nation governments for select installations in our sample to determine the extent to which these agreements include information on climate change adaptation, collaboration on climate change challenges, or cost-sharing related to climate change. Examples of these agreements include Status of Forces Agreements, Special Measures Agreements, Final Governing Standards, the Overseas Environmental Baseline Guidance Document, Technical Arrangements, memorandums of understanding, and other installation-level agreements between DOD and the host nation.

Further, we interviewed DOD officials from the regional service components, the sub-unified commands, and the installations in our sample to learn more about collaboration and cost-sharing related to climate change adaptation at the installation level. We also reviewed the Department of State's 2016 Treaties in Force for international agreements between the United States and host nations in our sample, which include information on climate change adaptation, and contacted the Supreme Audit Institutions of 28 countries to verify whether they had conducted any similar audit work.[10] We also interviewed Department of State officials from the Office of Security Negotiations and Agreements and the Office of Congressional and Public Affairs within the Bureau of Political-Military Affairs, and spoke with an official from the Office of the Legal Adviser to examine DOD's efforts to collaborate with host nations in our sample. In addition, we interviewed DOD officials from the Joint Staff; the headquarters of the Army, Navy, Marine Corps, and Air Force; the geographic combatant commands and their regional service components,

---

[10] Department of State, *Treaties in Force: A List of Treaties and Other International Agreements of the United States in Force on January 1, 2016.*

Appendix I: Scope and Methodology

the sub-unified commands, and the installations in our sample to learn about collaborative efforts between installation and host-nation officials and whether officials observed benefits to infrastructure used by DOD. To learn more about DOD's collaboration with host nations on cost-sharing activities and the extent to which these cost-sharing activities include information on climate change adaptation, we interviewed DOD officials from the Office of the Under Secretary of Defense for Acquisition, Technology, and Logistics and the Office of the Under Secretary of Defense for Policy. Also, we reviewed DOD Directive 4715.21, the 2014 Roadmap, both the 2010 and 2014 *Quadrennial Defense Reviews*, and the 2016 *Sustainability Plan* to determine DOD's goals for collaborating with external stakeholders—such as allies—on climate change adaptation. In addition, we reviewed past GAO work related to leading practices for collaboration.[11] We compared DOD's existing national- and installation-level collaboration practices with this body of DOD policy, guidance, and goals.

In order to select installations from which we gathered information for each of our objectives, we searched DOD's survey database to identify overseas locations and associated sites that reported climate change impacts. DOD has identified seven effects commonly associated with climate change: flooding due to storm surge, flooding due to non-storm surge events (e.g., rain, snow, sleet, ice, and river overflow), extreme temperatures (both hot and cold), wind, drought, wildfire, and changes in mean sea level. We developed a nongeneralizable sample of 45 installations that reported at least one of these seven effects; these installations were spread across 22 countries in each of the six geographic combatant commands' areas of responsibility.[12] To select locations to visit in person, we assessed each installation in our sample on the following five factors: the (1) number and type of climate change impacts reported for the installation; the (2) military service located at the installation; (3) the installation or site's plant replacement value based on DOD's 2015 Base Structure Report; (4) any planned military construction projects at the installation, as reflected in the geographic combatant commands' 2015 and 2016 Theatre Posture Plans; and (5) the country in which the installation and associated sites were located.

---

[11]*Managing for Results: Key Considerations for Implementing Interagency Collaborative Mechanisms*, GAO-12-1022 (Washington, D.C.: Sept. 27, 2012).

[12]The six geographic combatant commands are the U.S. Africa Command, U.S. Central Command, U.S. European Command, U.S. Northern Command, U.S. Pacific Command, and U.S. Southern Command.

Appendix I: Scope and Methodology

We also considered the severity of reported climate change impacts and the opportunity to observe impacts or adaptation actions in person. For example, we visited 14 installations within the U.S. European and Pacific Commands' areas of responsibility to observe both the physical impacts of extreme weather events and climate change on infrastructure and adaptation or resilience measures taken or planned at the installation level. During our site visits, we interviewed installation officials about observed impacts to infrastructure and collected key documentation describing impacts. Results from our nongeneralizable sample cannot be used to make inferences about all overseas DOD locations. However, the information from these installations provides valuable insights. Table 3 identifies the DOD installations we visited or contacted.

Table 3: Department of Defense (DOD) Installations We Visited or Contacted

Geographic combatant command	Location
U.S. Africa Command	Camp Lemonnier, Djibouti
	Royal Air Force Ascension Island, Saint Helena
U.S. Central Command	Masirah Air Base, Oman
	Naval Support Activity Bahrain, Bahrain
U.S. European Command	Army Garrison Stuttgart, Germany
	Army Garrison Rheinland-Pfalz (including Kaiserslautern and Baumholder), Germany
	Army Garrison Benelux (including Schinnen), Belgium, Netherlands, Luxembourg, and Germany
	Army Garrison Italy (Vicenza), Italy
	Army Garrison Livorno (Camp Darby), Italy
	Naval Support Activity Naples, Italy
	Naval Support Activity Souda Bay, Greece
	Naval Station Rota, Spain
	Naval Air Station Sigonella, Italy
	Royal Air Force Lakenheath, United Kingdom
	Royal Air Force Croughton, United Kingdom
	Royal Air Force Fairford, United Kingdom
	Incirlik Air Base, Turkey
	Lajes Field, Portugal
	Thule Air Base, Greenland
	Spangdahlem Air Base, Germany
	Ramstein Air Base, Germany
	Aviano Air Base, Italy
U.S. Northern Command	Atlantic Undersea Test and Evaluation Center, Bahamas

Appendix I: Scope and Methodology

Geographic combatant command	Location
U.S. Pacific Command	Army Garrison Kwajalein Atoll, Marshall Islands
	Army Garrison Japan (including Camp Zama and Kure), Japan
	Army Garrison Okinawa (Torii Station), Japan
	Army Garrison Daegu (including Camp Henry), South Korea
	Army Garrison Red Cloud (including Camp Casey), South Korea
	Army Garrison Humphreys, South Korea
	Naval Air Facility Atsugi, Japan
	Naval Air Facility Misawa, Japan
	Naval Support Facility Diego Garcia, British Indian Ocean Territory
	Commander Fleet Activities Yokosuka, Japan
	Commander Fleet Activities Sasebo, Japan
	Commander Fleet Activities Okinawa, Japan
	Commander Fleet Activities Chinhae, South Korea
	Yokota Air Base, Japan
	Misawa Air Base, Japan
	Kadena Air Base, Japan
	Osan Air Base, South Korea
	Kunsan Air Base, South Korea
	Marine Corps Air Station Iwakuni, Japan
	Marine Corps Base Camp Smedley D. Butler (including Camp Gonsalves, Camp Schwab, Camp Hansen, Camp Courtney, Camp Lester, Camp Foster, Camp Kinser, Camp McTureous, and Marine Corps Air Station Futenma), Japan
U.S. Southern Command	Soto Cano, Honduras
	Naval Station Guantanamo Bay, Cuba

Source: GAO analysis of DOD information. | GAO-18-206

By discussing potential sites for review with military service officials, reviewing relevant DOD reports, and reviewing relevant database characteristics, we determined that DOD's vulnerability assessment survey database was sufficiently reliable to use as part of our site selection methodology and to generate questions for data-gathering from sites visited or contacted. Also, by discussing the process by which the Office of the Secretary of Defense, military services, and Joint Staff selected survey sites, we determined that DOD's vulnerability assessment survey database was sufficiently reliable to assess the extent to which DOD effectively used the data to manage the operational and budgetary risks posed by weather effects associated with climate change. In addition, by reviewing relevant sites' data for any seeming outliers, we determined that DOD's Regionalized Sea Level Change Scenarios Database was sufficiently reliable to use as a source of data on which to base questions for sites visited or contacted and to illustrate cases in

Appendix I: Scope and Methodology

which installations may not be not using available data in their installation planning or project design efforts.

We conducted this performance audit from May 2016 to November 2017 in accordance with generally accepted government auditing standards. Those standards require that we plan and perform the audit to obtain sufficient, appropriate evidence to provide a reasonable basis for our findings and conclusions based on our audit objectives. We believe that the evidence obtained provides a reasonable basis for our findings and conclusions based on our audit objectives.

Appendix II: Comments from the Department of Defense

ASSISTANT SECRETARY OF DEFENSE
3400 DEFENSE PENTAGON
WASHINGTON, DC 20301-3400

ENERGY,
INSTALLATIONS,
AND ENVIRONMENT

OCT 20 2017

Mr. Brian J. Lepore
Director, Defense Capabilities and Management
U.S. Government Accountability Office
441 G Street, N.W.
Washington, DC 20548

Dear Mr. Lepore:

Thank you for the opportunity to comment on the Draft Report, GAO-18-206, "CLIMATE CHANGE ADAPTATION: DoD Needs to Better Incorporate Adaptation Into Planning and Collaboration at Overseas Installations," dated May 31, 2017 (GAO Code 100848). As requested during the exit briefing, we have provided technical comments on the Statement of Findings which was the basis for the narrative portion of the report directly to your lead analyst, Mr. Christopher Turner. Our comments on the recommendations are enclosed.

We appreciate your assessment of our efforts to prepare the Department of Defense (DoD) for the effects associated with a changing climate. The Department has always considered a wide-range of risks and has a proven record of planning and preparing for such threats. We continue to focus on ensuring our installations and infrastructure are fully resilient to a wide range of scenarios, while increasing lethality and supporting DoD operations worldwide.

The Department is currently reviewing guidance, including DoD Directive 4715.21, to focus on building resilience into our infrastructure. As we assess these policy documents, we continue to work across the Military Departments to incorporate resilience into planning and guidance.

The Department will continue to be prepared to conduct operations today and in the future, and will be prepared to address the effects of a changing climate on our threat assessments, resources, and readiness. I am committed to make our installations more resilient in support of our mission, our warfighters, and our communities.

Sincerely,

Lucian Niemeyer

Enclosure:
As stated

Appendix II: Comments from the Department of Defense

GAO DRAFT REPORT DATED MAY 31, 2017
GAO-18-206 (GAO CODE 100848)

"CLIMATE CHANGE ADAPTATION: DOD NEEDS TO BETTER INCORPORATE ADAPTATION INTO PLANNING AND COLLABORATION AT OVERSEAS INSTALLATIONS"

DEPARTMENT OF DEFENSE COMMENTS
TO THE GAO RECOMMENDATION

OVERALL: The draft report states in numerous cases a Department of Defense (DOD) position or policy that is neither current nor accurate. The report repeatedly cites DOD documents and threat reviews as though they are current as of July 2017. Using the term "According to DOD…" without a specific cite, attribution, or context is misleading and should not be included in a professional report. Ascribing infrastructure damage specifically to climate change impacts without taking into account the effects of extreme weather events is speculative at best and misleading (GAO has found in previous reports that it is not possible to link any individual weather event to climate change). The Department of Defense is in the process of updating a National Military Strategy and National Defense Strategy to focus resources on threats considered to be critical to our Nation's security and the security of our Allies.

RECOMMENDATION 1: GAO recommends that the Secretaries of the Army and Air Force, the Chief of Naval Operations, and the Commandant of the Marine Corps work with the Office of the Secretary of Defense (OSD) to issue a requirement to their installations to systematically track the costs associated with extreme weather events and climate change.

DoD RESPONSE: Non-Concur. Tracking impacts and costs associated with extreme weather events is important; however, the science of attributing these events to a changing climate is not supported by previous GAO reports. Currently, associating a single event to climate change is difficult and does not warrant the time and money expended in doing so.

RECOMMENDATION 2: GAO recommends that the Secretaries of the Army and Air Force, the Chief of Naval Operations, and the Commandant of the Marine Corps take steps to administer the Screening Level Vulnerability Assessment Survey (SLVAS), or a similar instrument, to all relevant locations.

DoD RESPONSE: Non-Concur. The SLVAS is an internally-developed survey instrument intended to provide an initial, broad-level screening of climate impacts across DoD. It does not provide quantitative data or account for mission criticality, and therefore is not a useful tool for long-term decision making. DoD will encourage the Military Departments to administer the SLVAS (or instrument they deem appropriate) as appropriate and as resources permit.

Appendix II: Comments from the Department of Defense

RECOMMENDATION 3: GAO recommends that the Secretaries of the Army and Air Force, the Chief of Naval Operations, and the Commandant of the Marine Corps implement DOD goals and plans by incorporating climate change adaptation into service-level guidance and required training for the development of installation-level plans, including master plans and natural resource plans at all locations.

DoD RESPONSE: Partially Concur. The Department is revising DoD Directive 4715.21 which will provide goals and requirements for all DoD Components.

RECOMMENDATION 4: GAO recommends that the Secretaries of the Army and Air Force, the Chief of Naval Operations, and the Commandant of the Marine Corps integrate climate change data and projections into the Unified Facilities Criteria (UFC) standards and periodically revise those standards based on any new projections, as appropriate.

DoD RESPONSE: Partially Concur. OSD will work with the Military Departments to evaluate the UFCs to determine where and when the use of climate data is appropriate, and modify and publish UFCs if warranted. For example, UFC 1-200-02 (December 2016), High Performance and Sustainable Building Requirements, includes general guidance to a designer in Section 3-7 to address climate risk without going into details of what data or projections to use. As climate projections are developed which inform engineering data, this guidance will allow DoD to direct designers to use it as appropriate on a project-by-project basis.

RECOMMENDATION 5: GAO recommends that the Secretary of Defense direct the Office of the Under Secretary of Defense (Policy), the geographic combatant commands, the sub-unified commands, the Secretaries of the Army and Air Force, the Chief of Naval Operations, and the Commandant of the Marine Corps to consider climate change adaptation as they develop DOD's position for future negotiations with host-nation governments on cost-sharing activities, when relevant or appropriate.

DoD RESPONSE: Partially Concur. The Department will work to review the processes and criteria governing host-nation cost sharing negotiations to strengthen or incorporate consideration of resilience when relevant or appropriate. Each bilateral agreement is different and individually negotiated, therefore a universal solution is not expected.

RECOMMENDATION 6: GAO recommends that the Secretary of Defense direct the Secretaries of the Army and Air Force, the Chief of Naval Operations, and the Commandant of the Marine Corps to issue guidance, as appropriate, that calls for more formal coordination mechanisms related to climate change adaptation, such as memorandums of understanding, between DOD installations and surrounding host-nation communities.

DoD RESPONSE: Partially Concur. The Department will work to review the guidance for establishing agreements between host-nation communities and DoD installations. If necessary, the Department will identify where it would be appropriate to incorporate consideration of resilience.

Appendix III: GAO Contact and Staff Acknowledgments

GAO Contact

Brian J. Lepore, (202) 512-4523 or leporeb@gao.gov

Staff Acknowledgments

In addition to the contact named above, Kristy Williams (Assistant Director), Laura Durland, Chris Turner, Leah English, Jane Eyre, Michael Fahy, Justin Fisher, Katherine Forsyth, J. Alfredo Gomez, Mae Jones, Marc Molino, Shahrzad Nikoo, Anne Stevens, and Joseph Thompson made key contributions to this report.

Related GAO Products

Climate Change: Improved Federal Coordination Could Facilitate Use of Forward-Looking Climate Information in Design Standards, Building Codes, and Certifications. GAO-17-3. Washington, D.C.: November 30, 2016.

Defense Infrastructure: DOD Efforts to Prevent and Mitigate Encroachment at Its Installations. GAO-17-86. Washington, D.C.: November 14, 2016.

Defense Facility Condition: Revised Guidance Needed to Improve Oversight of Assessments and Ratings. GAO-16-662. Washington, D.C.: June 23, 2016.

Military Training: DOD Met Annual Reporting Requirements in Its 2016 Sustainable Ranges Report. GAO-16-627. Washington, D.C.: June 15, 2016.

Defense Infrastructure: DOD Has Made Limited Progress in Assessing Foreign Encroachment Risks on Federally Managed Land. GAO-16-381R. Washington, D.C.: April 13, 2016.

Climate Information: A National System Could Help Federal, State, Local, and Private Sector Decision Makers Use Climate Information. GAO-16-37. Washington, D.C.: November 23, 2015.

Defense Infrastructure: Improvement in DOD Reporting and Cybersecurity Implementation Needed to Enhance Utility Resilience Planning. GAO-15-749. Washington, D.C.: July 23, 2015.

Military Training: DOD's Annual Sustainable Ranges Report Addressed Statutory Reporting Requirements. GAO-15-537. Washington, D.C.: June 17, 2015.

High-Risk Series: An Update. GAO-15-290. Washington, D.C.: February 11, 2015.

Defense Infrastructure: Risk Assessment Needed to Identify If Foreign Encroachment Threatens Test and Training Ranges. GAO-15-149. Washington, D.C.: December 16, 2014.

Standards for Internal Control in the Federal Government. GAO-14-704G. Washington, D.C.: September 10, 2014.

Related GAO Products

Climate Change Adaptation: DOD Can Improve Infrastructure Planning and Processes to Better Account for Potential Impacts. GAO-14-446. Washington, D.C.: May 30, 2014.

Military Training: DOD Met Annual Reporting Requirements for Its 2014 Sustainable Ranges Report. GAO-14-517. Washington, D.C.: May 9, 2014.

Extreme Weather Events: Limiting Federal Fiscal Exposure and Increasing the Nation's Resilience. GAO-14-364T. Washington, D.C.: February 12, 2014.

Climate Change: Energy Infrastructure Risks and Adaptation Efforts. GAO-14-74. Washington, D.C.: January 31, 2014.

Climate Change: Federal Efforts Under Way to Assess Water Infrastructure Vulnerabilities and Address Adaptation Challenges. GAO-14-23. Washington, D.C.: November 14, 2013.

Climate Change: State Should Further Improve Its Reporting on Financial Support to Developing Countries to Meet Future Requirements and Guidelines. GAO-13-829. Washington, D.C.: September 19, 2013.

Climate Change: Various Adaptation Efforts Are Under Way at Key Natural Resource Management Agencies. GAO-13-253. Washington, D.C.: May 31, 2013.

Climate Change: Future Federal Adaptation Efforts Could Better Support Local Infrastructure Decision Makers. GAO-13-242. Washington, D.C.: April 12, 2013.

High-Risk Series: An Update. GAO-13-283. Washington, D.C.: February 14, 2013.

International Climate Change Assessments: Federal Agencies Should Improve Reporting and Oversight of U.S. Funding. GAO-12-43. Washington, D.C.: November 17, 2011.

Climate Change Adaptation: Federal Efforts to Provide Information Could Help Government Decision Making. GAO-12-238T. Washington, D.C.: November 16, 2011.

GAO's Mission	The Government Accountability Office, the audit, evaluation, and investigative arm of Congress, exists to support Congress in meeting its constitutional responsibilities and to help improve the performance and accountability of the federal government for the American people. GAO examines the use of public funds; evaluates federal programs and policies; and provides analyses, recommendations, and other assistance to help Congress make informed oversight, policy, and funding decisions. GAO's commitment to good government is reflected in its core values of accountability, integrity, and reliability.
Obtaining Copies of GAO Reports and Testimony	The fastest and easiest way to obtain copies of GAO documents at no cost is through GAO's website (http://www.gao.gov). Each weekday afternoon, GAO posts on its website newly released reports, testimony, and correspondence. To have GAO e-mail you a list of newly posted products, go to http://www.gao.gov and select "E-mail Updates."
Order by Phone	The price of each GAO publication reflects GAO's actual cost of production and distribution and depends on the number of pages in the publication and whether the publication is printed in color or black and white. Pricing and ordering information is posted on GAO's website, http://www.gao.gov/ordering.htm. Place orders by calling (202) 512-6000, toll free (866) 801-7077, or TDD (202) 512-2537. Orders may be paid for using American Express, Discover Card, MasterCard, Visa, check, or money order. Call for additional information.
Connect with GAO	Connect with GAO on Facebook, Flickr, Twitter, and YouTube. Subscribe to our RSS Feeds or E-mail Updates. Listen to our Podcasts. Visit GAO on the web at www.gao.gov.
To Report Fraud, Waste, and Abuse in Federal Programs	Contact: Website: http://www.gao.gov/fraudnet/fraudnet.htm E-mail: fraudnet@gao.gov Automated answering system: (800) 424-5454 or (202) 512-7470
Congressional Relations	Orice Williams Brown, Managing Director, WilliamsO@gao.gov, (202) 512-4400, U.S. Government Accountability Office, 441 G Street NW, Room 7125, Washington, DC 20548
Public Affairs	Chuck Young, Managing Director, youngc1@gao.gov, (202) 512-4800 U.S. Government Accountability Office, 441 G Street NW, Room 7149 Washington, DC 20548
Strategic Planning and External Liaison	James-Christian Blockwood, Managing Director, spel@gao.gov, (202) 512-4707 U.S. Government Accountability Office, 441 G Street NW, Room 7814, Washington, DC 20548

Please Print on Recycled Paper.

www.ingramcontent.com/pod-product-compliance
Lightning Source LLC
Chambersburg PA
CBHW051157220526
45473CB00003B/806